T0243446

COYOTES AMONG US

COYOTES AMONG US

SECRETS OF THE CITY'S TOP PREDATOR

Stanley D. Gehrt, PhD, with Kerry Luft

Copyright © 2024 by Max McGraw Wildlife Foundation
All rights reserved.

No part of this book may be reproduced, or stored in a retrieval system, or transmitted in any form or by any means, electronic, mechanical, photocopying, recording, or otherwise, without express written permission of the publisher.

Some names and identifying details have been changed to protect the privacy of individuals.

Portions of this book are adapted from the website of the UCRP, urbancoyoteresearch.com. Chapter 1 is adapted from "Ghost Dogs and Their Unwitting Accomplices," published in the September 2021 issue of *Anthropology Now.*

Published by Flashpoint Books™, Seattle
www.flashpointbooks.com

Edited and designed by Girl Friday Productions
www.girlfridayproductions.com

Design: Paul Barrett
Project management: Mari Kesselring
Editorial production: Jaye Whitney Debber and Abi Pollokoff

Image credits: Cover: © Jaymi Heimbuch/Minden Pictures, Page ii: © Trevor Clark/Shutterstock, vi: Jeff Nelson, viii: Urban Coyote Research Project (UCRP), x (top): UCRP, x (middle): UCRP, x (bottom): UCRP, xii: Michael Novo, xiii: Alex Coombs, xv: Cory Arnold, xvii: Cory Arnold, xviii: Alex Coombs, 2: Courtesy RawPixel.com, 3: © Hody JW, Kays R (2018), 4: Courtesy Library of Congress, 5: Courtesy University of Alberta Library Prairie Postcard Collection, Wikimedia Commons, 6: Cory Arnold, 8: Cory Arnold, 9: UCRP, 10: UCRP, 12: UCRP, 13: Cory Arnold, 15: © Christine_Kohler/iStock, 17: Cory Arnold, 19: UCRP, 21: UCRP, 23: *Chicago Tribune*/TCA, 24: UCRP, 25: Jeff Nelson, 28: Jeff Nelson, 30: UCRP, 32: UCRP, 35: UCRP, 36: UCRP, 37: UCRP, 38: UCRP, 40: Jeff Nelson, 41: UCRP, 42: UCRP, 45 (left): UCRP, 45 (right): UCRP, 46: Cory Arnold, 48: UCRP, 49: UCRP, 50: Jeff Nelson, 52: © Debbie DiCarlo, 54: UCRP, 57: © Derek Audette/Dreamstime.com, 58: UCRP, 59: Cory Arnold, 61: © Jaymi Heimbuch, 62: Cory Arnold, 65: UCRP, 66: UCRP, 67: UCRP, 68: UCRP, 70: UCRP, 71: UCRP, 72: © mlharing/iStock, 74: UCRP, 76: Cory Arnold, 78: UCRP, 80: © Jaymi Heimbuch, 82: Cory Arnold, 84: Jeff Nelson, 85: UCRP, 86: UCRP, 87: UCRP, 88: © Chanawat Phadwichit/iStock, 90: Jeff Nelson, 92: UCRP, 95: © Wirestock/Dreamstime.com, 99: © Amina Akhtar, 103: © huntington/123rf.com, 106: Cory Arnold, 107: UCRP, 108: © ernestoeslava/Pixabay, 110: UCRP, 112: Cory Arnold, 114: Jeff Nelson, 116: Jeff Nelson, 118: UCRP, 120: UCRP, 122: NPS Photo / Emily Mesner, 126: Johanna Turner

ISBN (hardcover): 978-1-959411-23-9
ISBN (ebook): 978-1-959411-24-6

Library of Congress Control Number: 2023909311

Interior printed on FSC-certified paper from responsible sources.

First edition

For my parents, Jody and Earl Gehrt

CONTENTS

FOREWORD

For more than twenty years, the Max McGraw Wildlife Foundation has helped fund the work of Dr. Stan Gehrt and the dozens of graduate students and technicians who have made our property their home base while studying coyotes in the Chicago metropolitan area. For those who know of McGraw's work in conservation, the partnership may seem curious: an organization focused on the future of hunting, fishing, and land management—and more recently, the production of IMAX films and communications campaigns focused on conservation—sponsors research into a mammalian predator's adaptation to urban settings.

The philosophy behind this pairing is simple. At McGraw, we support individuals rather than programs, and Stan is an exceptional individual—as you will come to understand through this book.

Opposite: Research technician Lauren Ross setting up a remote camera.

Before he turned his attention to coyotes, Stan produced distinguished work on other species, such as bats and feral cats. This research, conducted while Stan was employed full-time at McGraw's Center for Wildlife Research, was so exceptional that we named him the center's director long ago. We continued that relationship after he joined the faculty at The Ohio State University, and when the proposal arose to study coyotes in conjunction with the Forest Preserves of Cook County, we were happy to sign on.

Stan is modest, so I must tell you that he is a pioneer in using satellite technology to track wildlife, as well as analyzing stable isotopes to determine a predator's diet. He is world-renowned for his work with coyotes, but his other studies on mammalian predators, such as raccoons on the North American prairies, are notable as well. His work has fostered a greater understanding of how people and wildlife interact and coexist, and in doing so it fulfills McGraw's purpose, laid out in its charter: to improve the world for humans.

Equally significant, Stan has built an amazing legacy through the hundreds of students he has taught and mentored over the years. His protégés are now doing their own groundbreaking research across North America and beyond, influenced heavily by Stan's high standards.

Finally, Stan is a splendid storyteller. I have sat spellbound in the audience many times over the years, listening to Stan explain his research and its implications to people who have almost no scientific background but who are nonetheless transfixed by his knowledge and passion for his subjects. It's no wonder that his semiannual presentations to our board and members are among the most eagerly anticipated events at McGraw.

Stan Gehrt has made McGraw an international leader in wildlife research. The story he tells in this book will show you how he did it and will make it clear why McGraw has proudly supported his work for more than two decades.

Charles S. Potter Jr.
President & CEO, Max McGraw Wildlife Foundation,
Dundee, Illinois
mcgraw.org

INTRODUCTION

There is a photograph taken at the beginning of the COVID-19 outbreak in 2020 that in many ways symbolizes the story of the coyote in North America. In it, a lone coyote lopes down Chicago's Michigan Avenue, a shopping mecca known as the Magnificent Mile. Yet in this photo, it is deserted. Except for the unseen photographer, there are no people anywhere—no shoppers, no pedestrians, no automobiles—merely a canyon between buildings and a skein of green traffic lights signaling to the coyote that the road ahead is open and clear.

The photograph is remarkable, a seemingly postapocalyptic view of urban America with a lone survivor. It is also a tribute to nature's ultimate survivor: the coyote.

Hounded, harassed, hazed, and hunted for generations, the coyote persists and even thrives. Though humans have nearly extirpated wolves and other large mammalian predators, coyotes have evaded all human efforts to wipe them out and have even expanded their range. Once confined to the American West, the coyote now lives in forty-nine of the fifty US states and has spread across lower Canada, into Mexico, and all the way to Panama. Sometime soon, coyotes will likely turn up in South America, the next step in a remarkable diaspora that has few, if any, equals in the animal kingdom.

Many coyotes still live in their original habitats, the scrub-brush

From residential neighborhoods to the most glamorous shopping districts, coyotes have made their homes in North America's cities.

deserts of the West and the open-sky country of the Great Plains. Yet as they moved east, coyotes displayed a stunning ability to adapt to new habitats. They are in the bayous of Louisiana, the river bottoms of Mississippi, the Florida swamps. They have traversed the Rockies and Appalachians, dodging cars, traps, and bullets. And now, they live among us.

Until the 1990s, it was rare to spot a coyote in an urban setting. Today, they live in virtually all of the United States' largest cities. In New York, they began filtering into the Bronx decades ago and now inhabit all five metropolitan boroughs. Los Angeles's healthy coyote population coexists with mountain lions. Seattle's coyotes traipse through the trendy Capitol Hill neighborhood, trotting past coffee shops and hanging out in nearby Volunteer Park. Name an urban center in North America, and it is probably home to a thriving coyote population.

With them, they bring fear. We blame coyotes for harassing other animals such as deer or livestock, for absconding with and eating beloved pets, for threatening people walking or bicycling through tranquil city parks. Was your garbage ransacked, or did your cat go missing? Pin it on the coyote.

The irony is that *we* are to blame. Rapid development and suburban expansions had the unintended consequence of creating excellent coyote habitats. Rodents, their favored prey, are everywhere in the city. So are rabbits, garbage, rotting produce, crusts and crumbs, and at times, food left out intentionally to feed wildlife. This smorgasbord combines with untold numbers of nooks, crannies, alleyways, and shrubbery for easy dens—and the coyotes have everything they need. They live in parking lots, arboretums, overgrown lots, and highway easements, and near water retention ponds. Even cemeteries bustle with thriving coyotes.

For many, seeing coyotes has become a somewhat unexpected part of urban life.

More than twenty years ago, in the hopes of better understanding these amazing creatures and their interaction with humans, we launched the Urban Coyote Research Project, a collaborative effort involving the Forest Preserve District of Cook County, Illinois; the Cook County Department of Animal and Rabies Control; The Ohio State University; and the Max McGraw Wildlife Foundation. We wanted to know whether coyotes could carry diseases that affect humans or pets, to gauge their impact on other wildlife such as deer, and to determine whether the coyotes could upset the forest preserve ecosystem. We also hoped that by learning more, we could minimize conflict between coyotes and humans.

We have learned all that and more, and we're still learning. The Urban Coyote Research Project is the longest-running coyote study in history, and by the end of 2022 we had tagged 1,433 coyotes, of which 683 were also outfitted with radio-tracking collars, and returned them to the spots where they were captured. Over time, we learned about their diet, their habits, their territories, and the threats posed to them and by them. As we will share in this book, we know now that coyotes are not nearly the pests they once were perceived to be, and that any effort to eradicate them is likely to fail.

It is therefore more practical for humans to consider ways to coexist with coyotes. There will inevitably be a small number of animals that become aggressive or threatening and must be removed by relocation or lethal methods. Still, over the two decades of our study, we have documented very few cases of coyotes biting humans. Often, these problem coyotes are created by well-meaning people who are fascinated by them and feed them. They want to help the coyotes, but feeding wild animals is never a good idea.

Since our study began, other coyote studies have launched in cities from coast to coast: San Francisco, Cleveland, Portland, Los Angeles, Atlanta, and Edmonton, Canada, among others. All of them have contributed significantly to our understanding of the coyote and to the

Coyotes downtown move along linear corridors, such as rivers, under the cover of night.

growing recognition that the coyote is worthy of our respect. By telling its story, we hope to illuminate the life of a magnificent creature that prefers to remain in the shadows.

Coyotes are the ghost dogs of the prairies and now of the big cities. Some may be within a few hundred yards of you even as you read this book. Seeing them should not be cause for alarm but rather a reminder that one of nature's most tenacious survivors will be living with us, and we with them, for some time to come. Coyotes are here to stay.

Behind the Image

Scoring an iconic wildlife image, like this one of Coyote #1288 with Chicago's Willis Tower (formerly Sears Tower) looming in the background, takes many things: expertise, hours of hard work, expensive equipment, and, in this case, a large dose of luck.

In early 2020, the research team and I began helping accomplished photographer Corey Arnold obtain images for a planned *National Geographic* feature story on urban wildlife. The story was to include our research, and it needed the high-quality images that the magazine expected. We began by using the GPS locations of select coyotes to guide Corey toward spots where he could set up his cameras.

Luckily, we had captured and radio-collared #1288, a young male, in early February, just before Corey's arrival. After moving south in the western suburbs, the coyote shifted his focus to the east. During his sixteen-mile journey, #1288 could have settled anywhere, or been hit by a car or train during his travels, but amazingly, he made his way to downtown Chicago.

Once we realized #1288 was sticking around the Chicago River, apparently with another coyote, we moved quickly to obtain a camera site. Corey began setting up his camera and three automated strobe lights, all orientated to frame the skyscrapers and eventually, hopefully, a coyote in the foreground. Yet almost immediately after setup, COVID-19 brought everything to a halt.

In fact, this camera setup was one of the last tasks we performed before we all isolated for the pandemic. While the city shut down, the camera kept monitoring and the coyotes kept doing their thing. Later that summer, Corey collected the equipment and inspected its memory card. The incredible image of #1288 was recorded on July 9, 2020, almost exactly four months after the camera was placed.

If the COVID-19 shutdown had happened earlier, if the camera and lights had malfunctioned or been stolen during that time, or if the coyote had continued his journey or died, this picture would not have been possible. In fact, #1288 was found dead a month after the image was taken. Many stars aligned to grace us with an amazing image of an amazing animal.

CHAPTER 1

Ghost Dogs of the City

If you are reading this book from almost any place in North America, in all likelihood you have been an unwitting accomplice to one of the most incredible wildlife stories of the past century: the story of the coyote (*Canis latrans*) and its unqualified success at conquering the continent—at least partially through its strange, paradoxical relationship with humans. To be clear, the consequence of the coyote's success is that you are—whether you are aware of it or not, and whether you are reading this from a rural farm, a residential subdivision, or even a downtown office—living with coyotes.

Opposite: Rail yards and backyards, parking lots and parks—all are frequented by the adaptable coyote.

The coyote's history has many layers that combine to make it so compelling, but two aspects in particular are the foundation of its incredible survival. First, in the early twentieth century—a period of extreme persecution of coyotes and conversion of the landscape to

An 1845 illustration from John J. Audubon's collection depicts male "prairie wolves."

primarily human use—the coyote dramatically expanded its distribution and abundance across the continent. Second, over the past twenty to thirty years, coyotes have become residents in virtually all metropolitan areas in the United States and Canada, a remarkable process that established them as an apex predator in urban systems built and occupied by *their* most dangerous predator: humans. This is where my research comes into play.

The coyote is a North American member of the Canidae family—which includes wolves, foxes, jackals, and, of course, domestic dogs. It's safe to say that coyotes have been around for a million years, though we lack fossil evidence to definitively state when the coyote became

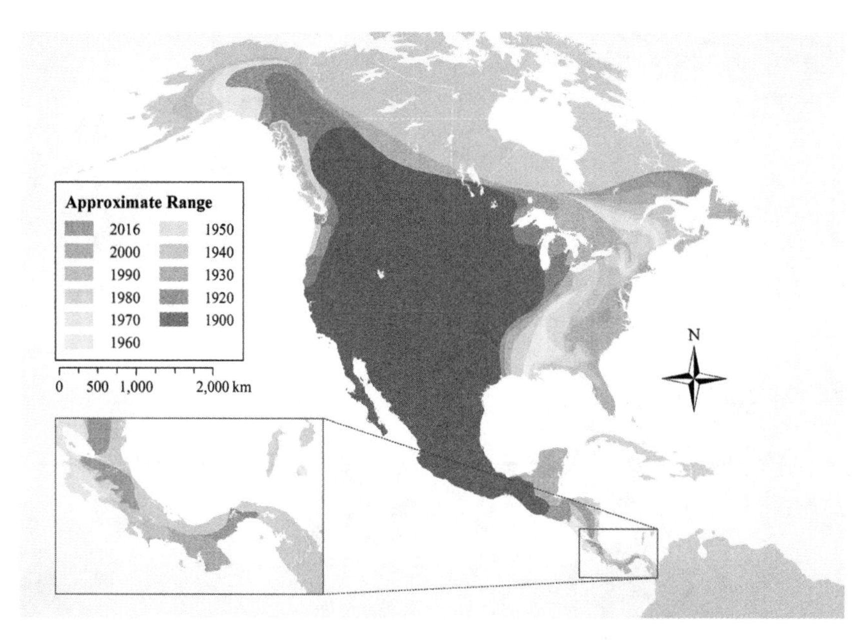

Coyotes have dramatically expanded their range since 1900.

the coyote, and which of the many possible canid ancestors produced it. At the time of European expansion across North America, from 1600 to 1850, the coyote's range was mostly restricted to plains and deserts west of the Mississippi and from the Canadian border to parts of Central America. On the open range of the West, the coyote already occupied the midsize carnivore niche, hunting mostly small prey and scavenging kills made by larger predators such as wolves and cougars, while at the same time avoiding those dangerous competitors. This lifestyle—hunting prey while avoiding larger predators—would serve the coyote well as one dangerous predator after another emerged. The coyote went from dodging dire wolves and saber-toothed cats in its earliest days to escaping mountain lions, gray wolves, and ultimately its most relentless predator of all, humans.

Little is known about how the first people to inhabit North America viewed coyotes. But coyotes play prominent roles in Western Native American lore and culture, suggesting some interaction and,

Ranger McEntire of the Malheur National Forest, Oregon, poses with coyote pelts in the winter of 1912 and 1913.

often, respect. Pioneer expansion by those with European ancestry made the landscape dangerous in a new way. Settlers swiftly targeted predators for removal and over time virtually wiped out wolves and mountain lions, among others. While these animals succumbed to persecution and were largely extirpated from their former ranges, coyotes responded by dramatically expanding their range, perhaps using the skills of evasion developed over thousands of years. Early accounts from Spanish explorers even described coyote-like animals in Costa Rica.

Still, at the turn of the twentieth century, coyotes were mostly found in the West, ranging all the way to California and the Pacific Northwest and down through Mexico. Maps of their progression show a slow creep across the Upper Midwest and a slightly faster expansion through Louisiana, Mississippi, and Alabama. By the end of the 1980s, they had reached the East Coast; within thirty years they were in all the lower forty-eight states plus Alaska, had spread across Canada, and had trickled into Central America. By 2013, they had crossed the Panama Canal. Will they continue south across the Darién Gap and into Colombia, establishing a beachhead in South America? We will have to wait and see, but as humans continue to alter the landscape, we facilitate the coyote's diaspora, building roads that serve as easy trails and raising livestock that sometimes provide a meal.

The amazing aspect of this tremendous range expansion is that it has been accomplished in the face of ongoing and intense human persecution. At no point have coyotes benefited from any sort of protection or conservation effort by state or federal agencies. Throughout most of the coyotes' range, people are allowed to hunt them for sport and remove them as predators. In most states, harvest regulations are more lenient for coyotes than for any other species. Unlike other game species, nearly all states allow year-round hunting of coyotes with no limits on the number that can be taken. Occasionally, coyote hunting

In this 1920s photograph, a priest attempts to train a coyote.

Benefits and Drawbacks of Urban Living for Coyotes

Benefits

- Food, in the form of natural prey and human-based products, is abundant all year and from year to year, compared to rural areas.
- Survival is much higher than in rural areas, especially for pups.
- Reproduction is higher, or at least litter sizes are larger than in rural areas.
- Urban coyotes are in good condition, often slightly larger or heavier than rural coyotes.

Drawbacks

- High densities mean more competition for space and mates.
- Roads are a challenge, and collisions with vehicles are the most common cause of mortality.
- Urban coyotes must limit their activity during the day to avoid us, thereby reducing their time for foraging and traveling.
- Urban coyotes often must travel farther and quicker than suburban or rural coyotes.
- Some diseases are more common in urban areas, such as heartworm and mange.
- Stress levels are slightly higher for urban coyotes than for those in natural areas.

contests, bounties, and other incentives appear to increase this persecution. Currently, between five hundred thousand and eight hundred thousand coyotes are killed in the United States each year through harvest or predator control measures. Of course, these are underestimates of actual deaths, because coyotes are killed in other ways, such as through recreational hunting and the removal of nuisance animals in urban areas, not to mention the tens of thousands killed by vehicular traffic. The total number of animals removed may approach one million in some years. Despite this pressure, and despite human efforts to control or remove them, the coyote not only persists but has more than doubled its range and increased in abundance. Put differently, after nearly two centuries of intensive "coyote control," there are more coyotes in more places on the continent today than at any point since European colonization. No other wildlife species can claim that level of invulnerability to human persecution.

As remarkable as that is, a more stunning aspect of the coyote's story is its amazing success populating all metropolitan areas in the United States and Canada. If coyotes had opposable digits, they would probably use them to thumb their noses at our efforts to exterminate them while they claim residence in our own backyards. But what does their perceived "success" in urban areas mean for us and our ever-evolving relationship with coyotes?

The emergence of the coyote in Chicago is emblematic of its presence in most major cities across the United States and Canada. Before the 1990s, coyotes were found only in the more remote areas of the Chicagoland region, and usually in low numbers. By the close of the twentieth century, their numbers appeared to have increased dramatically and coyotes began showing up where they had never been seen before, such as in suburban subdivisions. We don't know exactly why this happened, but it coincided with a crash in the animal fur market and a related falloff in trapping. Furbearer populations increased dramatically during that time as well.

A radio-collared coyote ignores a "no trespassing" sign while moving through the city.

As the twenty-first century approached, animal control agencies began fielding calls from concerned residents about coyotes, and many communities demanded the removal of some animals because of the perceived risk. Despite efforts to "depopulate" coyotes from the area, the coyotes persisted. This revealed the need for a better understanding of how the coyote population in the Chicago area was functioning and of the real risks coyotes posed to people and their pets.

At the time, the phenomenon of the urban coyote was relatively new, and little information existed upon which communities could rely in making coyote management decisions and, perhaps more important, calming the public's increasing fear. This need for basic information was the initial motivation for our Chicago research. Our project began in March 2000, when we captured and radio-collared a subadult female, Coyote #1. I still remember the excitement of capturing a coyote just a few miles from O'Hare International Airport, with airliners zooming overhead and thousands of cars roaring by only a

Coyote #115, also known as "Melonhead," during his only capture in February 2004.

few hundred yards away. Little did we know just how special that animal would become.

We began the study by assigning each coyote a functional—and admittedly uncreative—ID number corresponding to the order in which we captured them. Hence, Coyote #1 was the first coyote captured, and her second mate, a large handsome male, Coyote #115, was the 115th. At the end of 2022, our most recently captured animal was Coyote #1453. The rather boring numbering system not only helps to minimize anthropocentric influences in our science but also ensures that we maintain some degree of impartiality. We are observers of their lives, and we try not to influence what transpires, though this can be difficult.

Despite our efforts to remain detached, some individual coyotes do earn nicknames that stick, such as #115's moniker "Melonhead" because of his large head. The radio collars serve as windows into their lives, and the countless hours spent observing certain animals naturally produces a relationship with them, even if the coyotes are

Despite our fears, coyote attacks on humans are rare and frequently caused by well-meaning people who feed wild animals.

unaware of it. The collars also allow us to document the ends of their lives—we know we will inevitably record their deaths.

On my first night of following Coyote #1, I drove my truck and tracked her with radio telemetry as she took me on a convoluted journey across five subdivisions and a tollway. This trek ended in a patch of weeds, with my headlights shining on three men with dogs on leashes, completely unaware that a coyote was hiding only five yards away. In one night, that animal taught me the following:

- We were underestimating coyotes' ability to move through what we perceive as a challenging, urbanized landscape.

- We were grossly underestimating the coexistence already occurring between people and coyotes.
- We were likely underestimating the abundance of these animals in Cook County.
- And most important, I greatly underestimated the budget for this research! When we began, we had no idea of the thousands of miles we would drive tracking coyotes, nor the sheer number of animals we would need to track.

Coyote #1 and her mate taught us many other things over the decade we tracked them. Monogamous like all breeding pairs, they both lived exceptionally long lives of eleven years or more and raised at least thirty-eight offspring from seven litters. They spent every day living within a few yards of humans and their pets without conflict. One of their favorite hiding spots during the day was under a bush a couple yards from a post office, where hundreds of unsuspecting people walked past daily. To be fair, their pups sometimes destroyed Nerf footballs and stole chew bones from backyards, but these behaviors hardly warrant human persecution.

Over the course of our Chicago study, we have been able to observe a variety of behaviors and relationships that had not yet been reported. This was possible because we monitored many coyotes over many years in an urban system with a large number of resident coyotes and relatively high survival rates. Previous coyote studies had largely taken place in rural areas where hunting and predator control were prevalent. Research in these rural systems is less comprehensive because coyote territories are often temporarily vacant, space is available, and the number of coyotes competing for space is fewer. Further, it takes time to observe family dynamics, and most field studies rarely last longer than one or two coyote generations.

Coyotes have highly structured social systems, in which family groups, or packs, maintain exclusive territories that they defend from

Quiet, green, and seldom visited, cemeteries are favored locations for many urban coyotes.

other coyotes. As the population grows, more of the landscape is filled with these territories, and young (and sometimes older) coyotes leaving their packs will attempt to create a territory in a new area. Through this territorial system, the coyote population expanded across the Chicago region and into areas that had not previously experienced coyotes. Survival is relatively high and vacant territories are limited, so young coyotes are routinely forced to explore and attempt to exploit strange, novel areas—which they do quite well.

Our use of radio telemetry and GPS technology has shown that coyotes are capable of maintaining territories and raising litters in all parts of the Chicago area, even the most heavily developed regions where we originally thought it impossible. For example, possibly the most urban of our coyotes, #441, had a territory that encompassed all of downtown Chicago. Thus, she shared her territory with approximately 750,000 people, not counting the commuters who worked

Above: Dr. Stanley Gehrt tracks a coyote using radio telemetry in downtown Chicago. Page 15: An illustration from an 1898 book featuring North American animals shows coyotes hunting antelope as prairie dogs look on. We know that coyotes today hunt antelope fawns.

downtown. She lived there for at least five years without a conflict. Indeed, based on her location and the number of humans she shared space with, it's possible that she may have been the most "urban" coyote in the country.

During the day, #441 liked to rest on the island in the South Pond of the Lincoln Park Zoo or in a wooded spot at a marina about three-quarters of a mile away. Both areas were secluded from people, especially the island, which has no bridge and required her to swim or wade to get there.

She also hid in very public areas, such as under a bush near the entrance to the Alfred Caldwell Lily Pool, a small, gated area bordering the zoo, which is less than one hundred yards from busy Fullerton Avenue and less than a quarter mile from Lake Shore Drive, Chicago's chaotic expressway. Late one morning, I watched #441 for nearly an hour as she rested on the raised berm inside the main gate, where each person entering had to walk within twenty feet of her. Although

An Undeserved Reputation

Blame Mark Twain. In 1872, he published the semiautobiographical memoir *Roughing It*, relating his travels through the American West a decade earlier. In it, he wrote a description of a coyote that has pervaded common thought ever since:

> *The cayote is a long, slim, sick and sorry-looking skeleton, with a gray wolfskin stretched over it, a tolerably bushy tail that forever sags down with a despairing expression of forsakenness and misery, a furtive and evil eye, and a long, sharp face, with slightly lifted lip and exposed teeth. He has a general slinking expression all over. The cayote is a living, breathing allegory of Want. He is always hungry. He is always poor, out of luck and friendless. The meanest creatures despise him, and even the fleas would desert him for a velocipede. He is so spiritless and cowardly that even while his exposed teeth are pretending a threat, the rest of his face is apologizing for it. And he is so homely!—so scrawny, and ribby, and coarse-haired, and pitiful.*

At the time that Twain's account was published, the coyote was basically unknown east of the Mississippi River, even though one of the first in-depth descriptions of the species came from the great explorers Meriwether Lewis and William Clark. Within months of starting their expedition, on August 12, 1804, they spotted a small barking canine near present-day Onawa, Iowa. Determined to add it to their collection of new animal specimens, they set out after it and soon became the first—and certainly not the last—American hunters to be outfoxed by a coyote. Just over a month later, Clark and another expedition member each shot a coyote. With specimens in hand, and after more time observing what they called the "prairie wolf" in the wild, Lewis wrote a description:

The small woolf or burrowing dog of the prairies are the inhabitants almost invariably of the open plains; they usually associate in bands of ten or twelve sometimes more and burrow near some pass or place much frequented by game; not being able alone to take a deer or goat they are rarely ever found alone but hunt in bands; they frequently watch and seize their prey near their burrows; in these burrows they raise their young and to them they also resort when pursued; when a person approaches them they frequently bark, their note being precisely that of the small dog.

Further description of the coyote came some fifteen years later, in 1819, after the pioneering zoologist Thomas Say joined an expedition to the Rocky Mountains under the leadership of Army Major Stephen Harriman Long. While encamped near present-day Council Bluffs, Iowa, he wrote the first scientific description of the species and assigned it to the genus *Canis* with the species name *latrans*—which translates as "barking dog."

In 1948, around the same time the coyote began creeping across the continent, another coyote portrayal made an indelible mark on the American mind. Animator Chuck Jones, inspired by Twain's description, created Wile E. Coyote, a self-proclaimed "super genius" who became the Road Runner's scheming yet hapless foil. Amid ill-timed explosions, falling anvils, rocket-propelled roller skates, and other preposterous contraptions concocted by the Acme Corporation, Wile E. became a Saturday-morning icon and the prototype for many attitudes toward the coyote. In 1989, Jones told the *New York Times* that he imagined Wile E. as "a dissolute collie. . . . No one saw it more clearly than Mark Twain."

But Twain and Jones were wrong. Rather than a scoundrelly, skulking, bumbling villain, the coyote is among the most adaptable creatures on earth—a cunning survivor.

she was in the shade, anyone could have spotted her if they had only looked. Yet fifty or more people passed by without noticing the coyote hiding in plain sight.

To live in the city—or anywhere else—coyotes must avoid humans as much as possible. Most urban coyotes go about their daily lives largely unnoticed by people. To do this, they may hunker down during the day and move at night, sometimes within a few yards of human passersby. In fact, we have found that coyotes living in the most urbanized areas are exclusively nocturnal and travel farther distances within larger territories than more suburban coyotes. They also have a larger range on average—nearly three and a half square miles, compared to less than half a square mile for suburban counterparts. Part of this is because food may be more widely dispersed in the city, but it is more likely that urban coyotes roam farther because wide swaths of their territories are packed with people and other dangers.

Urban coyotes can learn human traffic patterns and know the safest times and locations to cross roads. They learn when and where humans are most active, and they scale their activities to avoid us. They spend a lot of time watching us and learning. Consequently, we largely coexist with them without even knowing it.

Opposite: Urban coyotes are creatures of the night, using the darkness to avoid human contact whenever possible.

Another important aspect of coyote success in urban areas is food. Our initial assumption was that coyotes did well in urban areas partially because of a reliance on human-associated food. In other words, we assumed that coyotes in cities were living off us, eating our trash and killing and feeding on our pets. Starting in 2012, we began analyzing coyote whiskers to characterize their individual diets. We did so because the traditional techniques that we used for more than a decade, such as fecal analysis, tended to underestimate the consumption of human-processed foods. The picture that emerged is that, much like us, coyotes are highly individualistic in their diets, even those living within the same packs and in the same areas.

13'-4"

Coyote Traditions, Legends, and Lore

Long before European settlers encountered the coyote, the Native people in the West had made it a seminal figure in their folklore. From the Pacific Northwest down through what would become Mexico and back up into the Great Plains, tribes viewed Coyote as a sly and greedy rapscallion, in some cases a friend to humans and in others a source of evil and trouble. Sometimes he is both. His English name derives from the Native American term ***coyotl***, meaning "trickster."

Like many mythological figures, Coyote is used as a means to explain natural phenomena. Some First Nations people believe that Coyote created the world, while others credit him with making humans. According to many cultures, Coyote gave humans their greatest gift: fire.

Coyote is especially significant to the Navajo, revered despite his faults and duplicity. Picaresque in nature, he mixes self-interest with altruism, always challenging the way things are. In some tales, Coyote is not too different from his pop-culture counterpart, Wile E. Coyote. He is male, anthropomorphic, and capable of concocting complex plots to advance his own interests—and often, those efforts backfire. A tale from the Southwest Pueblo Nation tells how Coyote is entrusted to deliver a pouch but is told not to open it. Intrigued, hungry, and unable to resist temptation, he unties the cords around the pouch and frees the stars, which fly off and populate the sky. As punishment, Coyote is inflicted with an unremitting toothache. The pain is so great he cannot sleep. He can only look at the night sky filled with the stars he set free, and mournfully, endlessly howl.

Many of the coyotes in our study are tagged as pups and returned to their den.

Surprising us, most coyotes we studied maintained diets dominated by natural foods, such as voles, mice, and rabbits, with only a minority relying heavily on human foods. Some whiskers we studied showed evidence of fruit, a reminder that some coyotes climb trees. We even have footage of a coyote slurping earthworms for twenty minutes as if they were strands of linguine—a previously unreported part of their diet. Basically, urban coyotes have a vast array of wild and human-associated foods available to them, and unlike for coyotes in rural systems, food abundance is maintained across seasons and years.

Other lines of evidence also support the conclusion that urban coyotes' dietary resources are not limited nor limiting. Chicago's coyotes, on average, exhibit excellent health and body condition. There is a small trend of increased size with urbanization among our population of coyotes—they tend to be heavier than rural animals.

Another indicator of the benefits of city life is litter size. Each spring, we enter the dens of our study animals to microchip and measure pups. We do this for many reasons, but a primary one is to record litter size. Coyotes can scale their litter size relative to available resources, so in times of plenty they may produce relatively large litters. We regularly record large litter sizes, which at times average more than eight pups per litter, and sometimes eleven or more. These numbers contrast with the usual five or six pups in a rural litter. Again, these lines of evidence reveal a picture of the metropolitan area as a hospitable refuge compared to rural areas. As a kid born and raised in a small Kansas town, I would have never guessed that an area with millions of people could be a coyote oasis.

There are, however, also costs to living in the urban world. Coyotes in the city core must travel farther and faster, and within a smaller window of time, to obtain resources. All coyotes must navigate roads, and for a transient, solitary coyote in a new part of town, any miscalculation could mean death. If a coyote suddenly becomes too obvious to people, by, for example, engaging in regular daytime activity, there will inevitably be a call to get rid of it. Although urban coyotes are relatively protected from hunting and trapping, human-related fatalities are still the most common causes of death, either unintentionally through vehicle collisions (by far their leading cause of mortality) or through intentional killings of so-called nuisance coyotes. A minority of these removals are the result of actual conflicts.

In extreme, very rare cases, a coyote can become aggressive—threatening or attacking pets or even people. More typically, a "nuisance" coyote is simply an animal observed at the wrong place at the wrong time. It could be active during the day instead of at night, or could be uncomfortably close to a school. A "nuisance" coyote could be an animal caught on a Ring camera in the middle of the night and posted to Facebook for the neighborhood to see. Or a group of coyotes may be considered nuisances without ever being seen, recognized only by their howls at night.

This extremely loose definition of a "nuisance" is unique among urban wildlife. Is any other species considered a nuisance and targeted for removal simply for being seen for a fleeting moment? It's a burden only coyotes carry.

In most cities, coyotes are the largest predator. They do attack pets rarely, and even more infrequently, people. Thus, urban coyotes represent a risk that was not present in most cities before coyotes expanded their range, and part of our research is to measure that risk. Each year, 1 to 4 percent of the coyotes we monitored in the Chicago area were removed as nuisances. In nearly all cases, the animal had not actually attacked or injured a pet or person but was becoming too obvious to people or may have conflicted with humans in other ways. For example, some coyotes are removed each year from airports, where there is understandably zero tolerance for disrupting flights. The large grasslands surrounding airports are unfortunately attractive for coyotes hunting rodents. Of the animals we have marked, only a handful have attacked pets, and none have attacked or threatened a person. In fact, our research has shown that many attacks blamed on coyotes actually involved dogs, and far more dogs than coyotes bite people in Cook County every year.

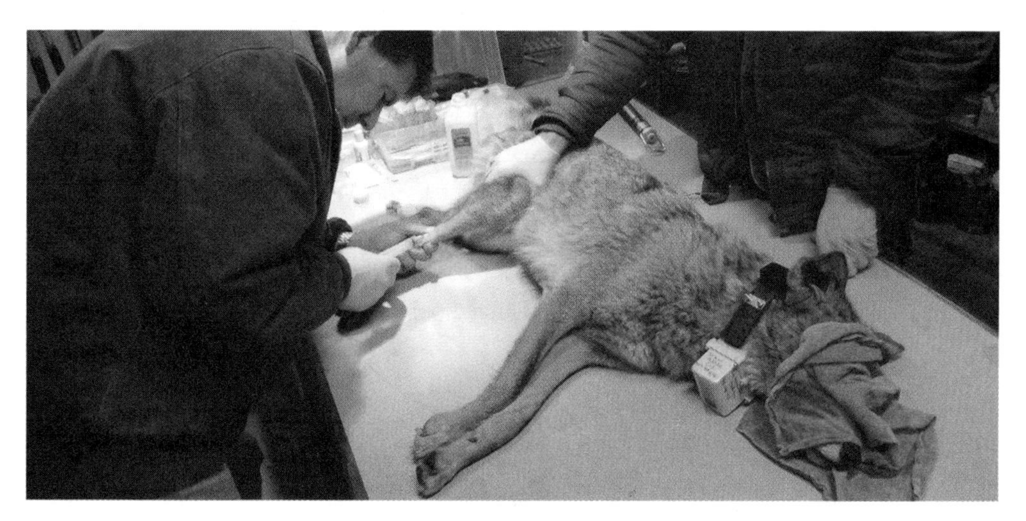

Small blood samples are collected for disease monitoring.

Coyote #441: The Most Urban Coyote

Coyote #441 is one of the most remarkable coyotes in our study. We captured her near Chicago's Lincoln Park Zoo on March 10, 2010. She was a subadult in excellent condition, weighing about 24 pounds. We outfitted her with a GPS collar, and we recorded her locations until November 2010, when her collar fell off as scheduled, allowing us to recover it.

But we were able to keep tabs on her via continuous camera surveys by Lincoln Park Zoo's Urban Wildlife Institute, identifying her by her numbered ear tags. She was one of the first coyotes we were able to study in downtown Chicago, and her movements shocked us. According to our data, she spent most of her time along Lake Shore Drive and within Lincoln Park, but she included several other city areas in her home range.

Multiple residents also spotted her on Chicago's downtown streets. One of our researchers even caught a glimpse of her while enjoying an evening in the city; he looked out a restaurant window just as #441 was trotting by. Despite her urban lifestyle, #441 kept a low profile, and she was never reported as a "problem" animal. She whelped at least two litters, based on her observed body condition during pupping season.

Her special radio collar gave us great insight into not only exactly where she traveled but also how. We programmed it to record her locations at specific intervals, allowing us to track her every movement. Interestingly, #441 almost always traveled on or near roads. We even documented her observing and learning traffic patterns and moving into or across intersections when the traffic signal changed, an ability that other resident coyotes have also displayed.

In the early evening hours of April 27, 2013, another researcher saw her trotting down a city alley, her yellow ear tags making her unmistakable. She was clearly lactating. Even though the researcher was walking his dog, #441 barely acknowledged them and just kept moving.

Coyote #441 spent much of her time in urban areas.

Later that year, the local ABC affiliate showed video of a coyote in a densely populated neighborhood on Chicago's North Side. Though her ear tags weren't visible, we believe it was indeed #441. A month after that, a citizen contacted our researchers and described a coyote resting in a Chicago park—and based on the description, it again was probably #441.

That was likely the last in-person sighting, but #441 would appear one last time. In January 2015, the *Chicago Tribune* published an article about the city's growing coyote problem and illustrated it with a picture of a coyote taken four years previously. We are almost certain it was #441, a true ghost dog of the city.

Coyote #748 protects his pups by laying on top of the den.

The characteristics of conflicts often vary based on the quirky nature of coyotes. Coyote #748, an alpha male (meaning he was an adult with a mate), occupied a territory encompassing Lake Shore Drive and some of Chicago's most iconic sites, such as the Field Museum, Soldier Field, and Willis Tower. He and his mate were "good" coyotes in that they avoided people and their pets at all costs. This changed in April 2014, when #748 suddenly became aggressive toward dogs, but in his own unique style. The pair had a newborn litter in a den at the top of a parking garage across from Soldier Field, a popular dog-walking spot along the shore of Lake Michigan. During the first two weeks after the litter was born, #748 would sneak down from the garage and, ignoring the poor dog owner, "attack" a dog to protect the den. Although there was a constant flow of dog walkers from early morning until late at night, #748 would target only one dog each evening and only between the hours of 6:30 and 8:30.

Some coyotes are larger than average. Coyote #1282 was a large adult male, weighing 38 pounds at capture.

Equally strange, he never actually injured a dog. He would jump on them and they would roll around together with much yelping in front of their terrified owner, but then he would trot away, leaving the dog covered in saliva but otherwise unharmed.

Many local residents wanted #748 to be killed, but because he had not posed a threat previously, we tried other methods first. For two nights, we chased him with noisemakers, and I shot at him with a paintball gun. This apparently worked, as #748 and his mate moved their litter to a different location away from dog walkers. At that point, #748 changed back into a "good" coyote in the eyes of the public.

This case also illustrates a common human quirk that likely contributes to perceptions of risk and trepidation regarding coyotes. People tend to exaggerate the size of animals, especially predators. No one ever reports encountering a "tiny" coyote, only "big" ones. I became aware of #748's switch to "dog attacker" only an hour or two after his first attack, because the dog owner Googled me and called my office when I happened to be working late. While he was walking his dog on a leash near the stadium, a "huge" radio-collared coyote came "out of nowhere" and jumped on his dog, who was luckily not injured. When the man described the coyote as weighing more than 100 pounds, I asked him how he estimated it to be that big. He said that the coyote was at least as large as his dog, a 110-pound mastiff. When we'd captured #748 the month before, he'd weighed a typical 29 pounds.

Somehow, with the animal right in front of him and even with his dog as a comparison, the owner managed to give 70 pounds of imaginary size to #748. Our ability to unconsciously inflate the size of animals we fear undoubtedly contributes to public concern. Although the actual risk of humans being attacked by coyotes is small, the *perceived* risk is often high.

Members of the public and officials often ask me: What good are coyotes? Why should people tolerate any risk, no matter how remote? Is there anything positive about coyotes in cities, or is the urban coyote story simply about managing risk? My answers to these questions may be surprising.

Predation is an important, even vital, function in ecosystems, and unfortunately, the lack of predators in urban systems results in the overpopulation of some prey species, often at the expense of habitats or property. For far too long, predation was absent or limited in our cities, allowing species such as geese, deer, and rodents to become artificially overabundant. When coyotes stepped onto the urban stage, they may have introduced predation to this severely altered ecosystem. Whether this is the case and to what extent became new research topics for us—and though we have some answers, we still have much more to learn. Each year brings new discoveries.

Through their perseverance and ongoing presence, coyotes are infusing themselves into our urban culture, as they originally did with

What's in a Name?

In North America, there are two basic pronunciations of *coyote*—the three-syllable "ki-oh-tee" and the two-syllable "ki-yote." Either is acceptable, as are minor variations, such as "ki-oh-tay." The term itself evolves from the Indigenous term *coyotl*. Spanish explorers turned that into "ki-oh-tay," and English-speaking settlers amended it further.

The basic differences in pronunciation today appear to be regional and possibly influenced by the Looney Tunes character Wile E. Coyote, pronounced with three syllables. In general, urban residents say "ki-oh-tee" and rural residents use "ki-yote." The two-syllable variation is most common in the Great Plains states such as Kansas, Nebraska, and the Dakotas, while along the coasts and in the East, "ki-oh-tee" is favored.

Native Americans and Western European settlers. Despite their best efforts to stay hidden, coyotes in the most urban areas have a difficult time avoiding people completely. As a result, modern urbanites who encounter them have developed their own "coyote stories." When we began our research two decades ago, urban coyote stories were rare, but now they are common and even transcend continental boundaries.

One of my favorite anecdotes comes courtesy of Coyote #441, the adult female whose territory comprised downtown Chicago. In 2010, I received an email from a person from Switzerland who traveled to Chicago each year for a week of business meetings. He wondered whether he had encountered one of our study animals on his most recent trip. When he stayed in Chicago, his routine was to attend meetings all day and then run at night in Grant Park. As he jogged one warm summer evening, he was surprised by a doglike animal passing him from behind on the path. It was not leashed but it was

All coyotes have a single molt, when they replace their coat, during late spring and early summer.

wearing a strange collar. The animal gave him a quick glance but never broke its effortless trot as it continued down the path. It happened so quickly, the business traveler wasn't sure whether it was a coyote or a strange dog. However, as he continued his jog around the park, he kept an eye out for the animal. Sure enough, before he had completed his lap, it came from behind again and, like before, barely acknowledged him as it casually lapped him, passing by a few inches from his leg. He thought the animal shot him a glance that was mildly approving of his progress, and then it was gone.

When I responded to his email that, yes, this was Coyote #441, and she regularly used Grant Park, he was thrilled, using many exclamation points!!! He described his encounter with one of the famous "Chicago coyotes" in Grant Park as being easily the most memorable experience from all his business trips. He said he would remember it forever.

This anecdote makes me smile each time I share it. And it is worth noting that it is one of thousands of coyote encounters that take place each year that are not conflicts; rather, these meetings of human and coyote are the spice of life—memorable moments that are never reported in the media, unlike the rare cases of an attack on a dog. Much like their ecological effects on the urban ecosystem, coyotes are likely affecting human culture in subtle ways that have not been fully recognized.

Even if you do not have your own coyote story, there's a high degree of certainty you are living with coyotes. If you live or work within a metropolitan area, at some point you have passed within a few yards of a coyote. You may pass them regularly on foot or by car. When you use a park, visit a cemetery, run an errand in a shopping mall, or play a round of golf, undoubtedly there is a coyote watching and learning from you. As you commute to work, a coyote is near the road or rail line, avoiding you.

It is through your activity that you reinforce or change their behavior, and as a result, you are playing a role in one of the most amazing wildlife stories in North America. The coyote's ability to live in an urban area and effectively coexist with us relies on its ability to avoid us. Just as coyotes learn from and respond to us, we can also learn from coyotes if we are willing. To many First People, coyotes were known for their clever, sometimes mischievous ways. In early American pop culture, coyotes were "wily" and secretive. They continue to outsmart us, whether we are the scientists attempting to understand them or the people trying to exterminate them. Every day they teach us that there is much we don't understand about this world, even in our own backyards.

CHAPTER 2

Studying Coyotes

For the past twenty-two years, I have had the privilege of developing and supervising the Urban Coyote Research Project, the largest study of coyotes to date within one of the largest urban centers in North America. Over the years, we have used various types of technology to peer into the lives of these animals—lives that remain largely hidden despite their existence within a landscape containing nine million people. Even with the advantages of radio telemetry, GPS satellites, remote cameras, chemical analysis of tissue for diet, and the latest genetic tools, it never ceases to amaze me how difficult it is to study coyotes. In many ways, they are as mysterious to me as when we started.

Opposite: Coyote pups are irresistible balls of fur, but that stage of their life doesn't last very long.

The Chicago region is one of the most heavily urbanized areas on the North American continent. It comprises all or part of six northeastern Illinois counties and sprawls north into Wisconsin and south and east into Indiana. Hundreds of municipalities dot the landscape. It is a major transportation hub, with railroad lines and interstates crisscrossing the region, and O'Hare International Airport, once the world's busiest, remains an important focus of air travel and shipping. What in the world are coyotes doing here?

Though the natural savannas, woodlands, grasslands, and wetlands that originally covered this landscape are mostly gone—first falling to the plow, then being covered by urban and suburban sprawl—fragments of habitat remain. In Cook County, which includes the city of Chicago, more than 10 percent of the land is protected from development, mainly through the establishment of a network of forest preserves. Much of our fieldwork takes place in or near those preserves, especially the ones in the northwestern portion of the region—including O'Hare.

Conducting research on urban coyotes is challenging. The two most daunting and difficult aspects are, first, capturing coyotes in an urban landscape with millions of people and, second, trying to track them once they are released.

We use humane traps to capture coyotes. While in hand, the

Coyote #748 stands on a parking garage across from Soldier Field.

coyotes are evaluated for overall health. We collect basic biometric data, such as weight and other physical measurements, and draw blood for later disease analysis. Each coyote gets ear tags (yellow for females, red for males). The tags have unique numbers that further help us identify individuals in the field. Most captured coyotes also are outfitted with radio collars, allowing us to track them with truck-mounted antennae, via direct observation, and, in the case of some more advanced collars, by using GPS data. The GPS radios are located by satellite, allowing us to track these coyotes remotely. This helps us determine their home range, where they live and regularly travel. Radio collars also allow us to document how long coyotes survive and their causes of death.

Coyotes are some of the most difficult animals to capture. To do so successfully, we have to outwit them and overcome their amazing sensory capabilities while taking advantage of their inherent curiosity. We use many of the same techniques as fur trappers. We can't leave our scent at trap sites—even a drop of sweat may be enough to deter a coyote. A stone overturned on a trail may attract their curiosity or may cause them to leave. They have very sensitive feet and can feel a shift in the soil where a trap is buried. Once traps are set, we hold our breath, hoping that we don't end up educating a coyote. If we do, that coyote will likely never fall for that trap again. After taking hours to carefully set a trap in the ground, meticulously disguising it, and being cautious not to leave our scent, the worst thing is to check it the next morning and find the trap carefully dug up—as if the coyote wanted to show you how silly you are. And, if you are extra unlucky, the coyote will make its point by leaving a scat sitting on top of the hapless trap. What other animal has such a snarky sense of humor?

Once we are lucky enough to capture a coyote, we place their collars and return them to the exact location where they were trapped. Then we track them day and night—in many cases we are out from dusk to dawn, focusing on a particular group of coyotes and figuring

Coyotes #1 and #115: The Beginning

On March 22, 2000, we captured Coyote #1—the first one for the project, and over time she became the signature coyote of our research. When first captured, she was a solitary transient, just one year old. She weighed about 29 pounds, and though on the small side, she was in excellent health. In coyote years, she was a teenager.

We tracked her over portions of five municipalities for the next nine months, as she floated across the landscape seeking her own territory. In early 2001, she settled down with an uncollared male and started a pack. Upon the death of her first mate, likely due to a vehicle collision, she mated with another male that we did capture, #115, whom we dubbed "Melonhead." We captured him on February 18, 2004, at the peak of mating season. Coyote #115 was large and in excellent condition, weighing in at 40 pounds. They remained together for the rest of her life.

After months of trying, we recaptured #1 on April 12, 2004. By then, she was a mature, pregnant female. Again, she was in excellent health, weighing in at 37.5 pounds. She would live six more years, and we were fortunate to follow her every year. We also tracked many of her offspring (we were able to microchip or radio-collar thirty-eight of her pups with #115, covering six litters). Remarkably, Coyote #1 died of natural causes at the advanced age of eleven in 2010.

She was obviously very street smart, given that her territory covered many busy roads. As a subadult (between one and two years old), she was seen crossing eight lanes of traffic on Interstate Highway 290, and many researchers watched her cross roads regularly at night.

Her relationship with #115 was interesting to observe. At times they were inseparable, and at others they would take short breaks from each other. Still, they continuously defended the same territory together. Over that time, they had at least seven litters together, with both parents helping to raise the young. We never received complaints about either coyote, even though they lived in a heavily developed area not far from O'Hare International Airport.

Coyote #115 was quite old the last time we saw him, his age evident in his gait, his coat, and his eyes. He had outlived the life of his radio collar, so we aren't sure what happened to him. Efforts to continue tracking him failed. Our last encounter was when we found pups in his territory in 2012. Genetic analysis of the pups revealed that Melonhead had found at least one more mate after #1's death.

You may have seen Coyote #115: He appeared in the Canadian Broadcasting Corporation's misleadingly titled documentary *Meet the Coywolf*, which aired on PBS in early 2014.

Coyote #1 during capture.

Above: Our tracking collars are programmed to fall off after a certain period, allowing us to recover them, download their data, and retrofit them for another coyote. Opposite: Some coyotes in our study are equipped with a Crittercam, which stays on for three days or less.

out their location every hour or two. This gives us a good indication of the extent of their normal range. Most of the latest collars are programmed to fall off after a set period of approximately a year, allowing us to recover them, download their data, and then reuse them on other coyotes.

Radio tracking is another tremendous challenge for our project. Most of the technicians' time on the project is spent driving on Chicago-area roads, fighting through traffic and listening for beeps indicating a radio-collared coyote is nearby. Radio interference can be horrible in some areas. In the early years of the study, we focused our fieldwork within a three-hundred-square-mile area. However, we have now expanded the fieldwork across most of Cook County. Obviously, we must follow where the coyotes take us. Night after night, hour after hour, a technician is on the road in a tracking vehicle, seeking radio-collared coyotes.

We've even outfitted some coyotes with video cameras strapped under their chins. Part of a collaboration with National Geographic, these "Crittercams" give us great insight into coyote behavior—how they eat, travel, and avoid human conflicts. For example, we've seen coyotes carefully pluck dead birds they've found before eating them, among other gustatory surprises.

Some of their antics can be downright comical. In one instance, a female coyote approached a stationary blob of fur. It was her mate, sleeping. She began nudging and nosing him, like a spouse prodding her mate to remind him that there was work to be done. He slowly uncoiled and peered back at her, seemingly asking why he couldn't just enjoy a bit of quiet time. Ultimately, he gave in and stood up, and they trotted away together.

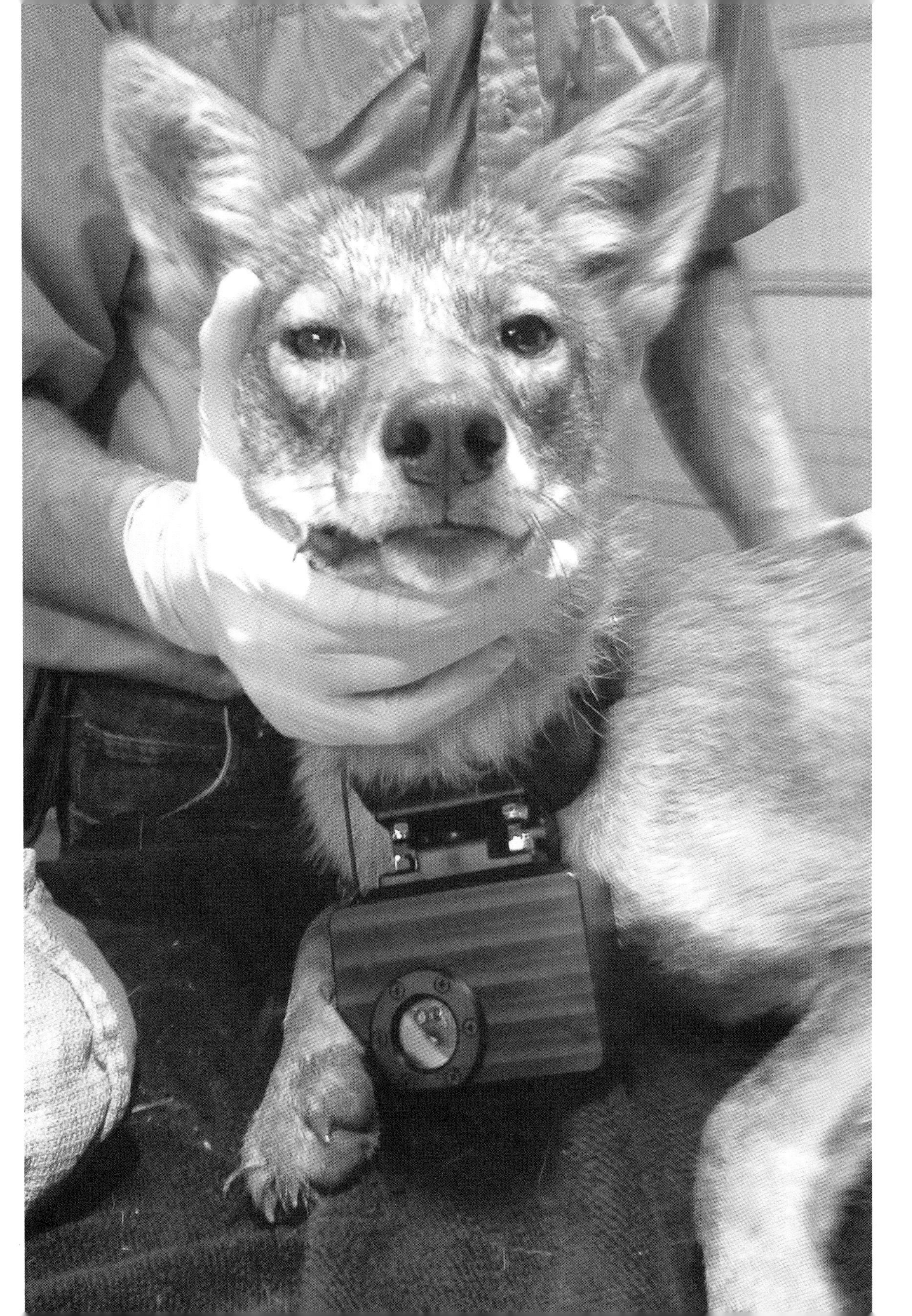

Besides trapping and releasing adults, we also locate and mark coyote pups in the spring pupping season, a process that begins with a den, some hope, and a flashlight. It's not as easy as simply finding a coyote den and diving in. Coyotes use multiple dens, so you must find the right den at the right time, hopefully within two to four weeks after the mother gives birth. In the Chicago area, that occurs between late April and early May.

Our researchers look for telltale signs. For example, as coyotes burrow into the ground they fling dirt behind them. If fresh, these piles of soil, called tailings, are a good indication that pups might be inside the den. If unsure, the researcher can brush away some of the soil. If the grass beneath the soil is green, the tailings are fresh.

Then comes the smell test. Coyotes are clean, so any moldy or rotting smell means the den is no longer active. What we hope to sniff is a slight, distinctive musk that's surely present but not overpowering. You'll know it when you smell it.

Researchers must round up coyote pups by going into the den headfirst.

At that point, it's time to dive in. One member of our team starts burrowing into the hole headfirst. They may hold a cell phone in front of them, in video mode and with flashlight on. This helps the researcher peer into the gloom and around bends in the den's tunnel, always looking for the reflection of small, bright eyes or even just a bit of movement. If the pups are on the older side, they might growl a bit.

If that expedition doesn't conclusively find pups, the researcher will start probing with a long stick. Because coyote dens can go back as far as ten or twelve feet, team members sometimes double up in the tunnel, hanging on to the ankle of the person in front. Once we even had three researchers in a tunnel, with only a pair of feet left poking out of the entrance.

Coyotes are elusive, even as pups. One of our teams once spelunked through a den for almost an hour and found nothing. As they prepared to leave, a chubby pup waddled out and looked at the researchers as if to say, "You missed me!"

Once we've located the pups, it's time to bring them out into the world. If a den is especially long and complex, we may dig down from above—but that's a last resort, as we want to disturb the den and pups as little as possible. More often, a researcher hollows out the opening to the den enough to let their shoulders pass through and then begins belly-crawling with a leather-gloved hand reaching forward. Though the pups usually aren't a danger, research associate Shane McKenzie once felt something nipping at his hand as he probed. It was the *mother* coyote. Shane didn't know she was there because we had not collared her previously. Fortunately, she chose not to attack.

We try to grab the pups by the scruff, as their mother would, but sometimes we must snag a leg instead. Then we gently ease them through the tunnel and into the hands of our waiting technicians.

They are irresistible—tiny squirming balls of fur, in a riotous spectrum of colors: tan, gray, black, brindle. Early on, their eyes are closed and their ears clamped flat to their heads. At that point, the pups are

Capturing Adults

As you might expect, it is far more challenging to capture, collar, and release an adult coyote. It takes teamwork, experience, and a great deal of caution so that neither our researchers nor the animal is injured.

First, we look for signs of coyotes, such as tracks and scat. We set up traps that are modified to minimize injuries, including jawed foothold traps that are either padded or do not close all the way, or a cable that is set up to loop around the animal's neck and restrain it without choking. Sometimes in more natural settings we use bait, in the form of deer roadkill, organs from deer culled by forest preserve sharpshooters, or pheasant carcasses from a local hunting preserve. To prevent the coyotes from dragging the bait away, we stake down deer legs. The other forms of bait are usually consumed on the spot.

Then, we frequently visit the trap sites so we can replenish bait if necessary, or recover any animals as quickly as possible. If a

coyote has been caught, two research technicians work together to swiftly secure it. People are often surprised to learn that captured coyotes are usually quite docile or passive in the traps as we try to remove them. They seem to know that resisting is futile.

Next, the coyote is placed in a cage and transported to a laboratory at the Max McGraw Wildlife Foundation or the Cook County Forest Preserve District Wildlife Management facility. The coyotes handle the travel very well, much like a pet dog that is accustomed to traveling in its crate.

Upon arrival at the lab, the technicians work fast. They inject the coyote with a fast-acting sedative, and once it is unconscious, they take samples of blood, fur, whiskers, and scat, and outfit the coyote with ear tags and, in most cases, radio-tracking collars. They also make a general assessment of the coyote's health. The goal is to have the coyote unconscious in five minutes and the entire procedure done in twenty.

After four or five hours of recovery, we return the animals to the same area where they were captured. If the coyote was captured in a busy area, we wait until dark to return it in order to avoid periods of heavy traffic or pedestrians. We open the cage and back away. Some coyotes are timid and hesitant to leave the cage. One stayed inside for fifteen minutes. Most dart away, disappearing into the night.

extremely docile because they're sleeping most of the day. They sag in the hand, gently mewling.

Coyote pups are born blind and deaf. Ideally, we find the pups when they're about two and a half weeks old. They weigh about a pound then, and their eyes are open, their ears are erect, and they're beginning to teethe. In our hands, they squirm and fuss for a bit but calm quickly. Their fur is fine and velvety; it thickens and gets coarser as they grow. Some burrow into the blankets our team uses to keep them warm, while others hold back. We weigh them, take hair samples, quickly draw blood, and inject the pups with a microchip like the ones used to keep track of pet dogs and cats.

Opposite: Docile and calm, a coyote pup is ready to be examined and tagged—a process that takes mere minutes.

These procedures are swift and relatively painless; the pups usually don't even yelp. By doing this, we can study the pups' genetics and identify them later in their life, either when we capture them as adults or, sadly, find their bodies. Then we carefully return the pups to the den and clean up any traces of our presence. The whole procedure is done as quickly as possible, usually within ninety minutes, for we know through our radio telemetry that their parents are close by, watching over their offspring.

Our project is wide reaching. Between the beginning of our research and the end of 2022, we tagged 1,433 coyotes. Of those coyotes we captured, we also fitted 683 with radio collars and recorded more than 120,000 locations, allowing us unprecedented insight into their hidden lives.

Although this study is focused in Cook County, Illinois, what we have learned about coyotes and humans living together is indicative of many metropolitan areas. The aim of this research is to take the first steps toward coyote management: public education. Through our efforts, people can better understand the difference between true threats and coexistence with coyotes. Coyotes react to us, and we can foster mutual respect or a lack thereof through the cues we send to

Coyote Personalities

Our most recent research efforts focus on a current hot topic in urban wildlife ecology: Do novel processes in urban ecosystems, such as traffic, human foods, noise, and so on, "select" for certain types of personalities in wildlife? Could this be happening with coyotes? We know that coyotes have individual personalities. We see a variety of personalities while handling pups at the den, especially by five weeks of age. Within a litter, some pups snarl, some attempt to flee, and others passively allow us to handle them. But we don't know how much of adult coyote behavior is the result of an ingrained personality or of learning from experience. We might describe a coyote that crosses a road or takes a bag of food with human scent on it as being bold. But is that because it has a bold personality, or because it learned from past experience?

Two PhD graduate students from The Ohio State University in our lab used different approaches to attempt to answer this question. Katie Robertson attempted to measure the behavior of resident coyotes within their home ranges by using novel object tests. To do this, she put either a child's ring toy or a yard gnome in a frequently used area of a radio-collared coyote's home range and monitored their response with hidden cameras.

We loved watching videos of the coyotes reacting to the objects! It was especially amusing to see them shy away from the yard gnome, clearly not knowing what to make of it. One coyote exhibited the ultimate bold behavior by lifting his leg and marking the gnome, which was quite funny until Katie realized she would have to wash it.

After comparing her data across individuals living in different levels of urbanization, Katie found a trend toward more bold visits by coyotes in highly urbanized parts of Cook County. However, most coyotes were reluctant to approach the objects.

At the same time, graduate student Ashley Wurth recorded coyote behavior in a more controlled situation than Katie's more natural

experiment. When we captured coyotes as part of our overall study, Ashley recorded their behaviors at different stages of the process, such as in a trap, when being removed from the trap, when in the handling cage, and upon release.

Over the four years of this study (2014–2018), we captured, radio-collared, and recorded behavior for 137 coyotes across Cook County, and we developed personality profiles for each individual by combining the various behavioral measurements she recorded.

A primary finding was that 75 percent of the coyotes possessed shy personalities, and bold coyotes were in the minority. Yet, when Ashley compared results across different levels of urbanization, she also found most bold animals lived in more urban settings—a result consistent with Katie's work. Two very different approaches yielded similar results. Thus, we believe the coyote population in Cook County is characterized by shy personalities, but that bold coyotes exist and tend to occur more frequently in heavily developed areas. This pattern suggests that the city may be "selecting" for bold coyotes, but time will tell as to whether bold coyotes become more common in the population.

them. In this research, our primary interest is the observation of natural behaviors versus experimental manipulations.

Over the years, we have come to conclude that we consistently underestimate coyotes and their ability to adjust and adapt. They push the boundaries of what we might consider constraints, constantly finding new ways to humble us and confound our perceptions of their limitations. It is an animal worthy of our respect, not scorn.

Some of the characteristics revealed by our research are hailed as admirable qualities in humans. One of the most surprising findings, in a research project full of surprises, was what we uncovered when we explored coyote mating through genetics. This exploration also revealed more mysteries.

Monogamy is rare among mammals. Roughly 90 percent of mammal species mate with multiple partners, with the obvious benefit of increasing the genetic variety of their offspring. Bucking this trend is the Canidae, the family of wolves, coyotes, foxes, and jackals, in which virtually all species are socially monogamous, except for domestic dogs.

Alpha pairs often travel together, although you may only see one. Here, an alpha pair travels together under cover of night, with one mate a mere shadow of the other.

Yet social monogamy does not ensure genetic monogamy. Once geneticists began poking around apparently monogamous species, they found true genetic monogamy to be quite rare. This makes sense when you consider that spreading one's genes among several mates is the equivalent of not putting all one's eggs into one basket. Given the relatively high densities of coyotes and their high survival rate in parts of Cook County, we assumed we would also uncover "cheating" among our coyotes once we put them to the genetic test.

Instead, ours is the only study to report 100 percent genetic monogamy among canid species, though they all exhibit social monogamy. For example, studies reported nearly 40 percent of offspring born to arctic fox pairs were from "extra" fathers. Thirteen percent of wolf pups were the result of "cheating," as were 40 to 80 percent of red fox offspring.

These genetic results lead to more questions. Why do coyotes not "cheat" when it is universal among other species? Why are pair bonds maintained for life, or even maintained during the year when no reproductive behavior is occurring? And how do coyotes select their mates (one of the questions most fascinating to me)? Because they apparently mate for life, a young coyote's choice of a mate may be one of the most important decisions it makes if the goal is to ensure their genetic legacy. My thoughts return to Coyote #1 and her longtime mate, Melonhead. Was it truly a random choice based on nothing more than timing? Or do coyotes have a way of assessing the fitness of possible mates? We don't know.

Their fidelity is truly remarkable since the opportunity for taking multiple partners would appear to be substantial in the Cook County population. Solitary coyotes are omnipresent, furtively moving through territories, looking for opportunities, effectively providing a member of a pair an opportunity to stray without leaving home. Yet, as of 2023, all alpha coyotes in our study—male and female—have resisted these temptations.

A mated coyote pair, both sporting radio collars.

Fidelity involves cooperation in raising and protecting young, which allows the pair to have large litters and increased reproductive success. Also, an alpha pair can more easily defend a territory; in fact, we have observed some packs lose part or all of their territory upon the death of an alpha. So there are benefits to monogamy, but it is still impressive to see no cheating among our coyotes.

Nor have we seen evidence of serious fighting among coyotes. Only one animal in our study might have died from injuries caused by other coyotes. By contrast, wolves often kill each other. In Yellowstone National Park, where wolves are protected from hunting, intraspecific fights are the wolves' leading cause of death.

Each of our discoveries has led to new avenues of research, propelled by advances in technology and methodology. Two decades might seem to be a long time to study a specific animal in a defined region, but we have so much more to learn. Twenty years from now, we will know a great deal more about how coyotes have adapted to living among us, and whether we have learned to live with them.

Coyote #466: A Coat of Many Colors

Coyote #466 passed his unique coloration onto his offspring.

We captured Coyote #466 on May 12, 2010, in the Crabtree Nature Center near Barrington, Illinois. He was an adult in excellent condition and weighed 39.7 pounds—one of the largest coyotes we have tagged. But what really made him exceptional was his coat, a hodgepodge of dark blacks, browns, and grays. He was a brindle coyote, one of only a few tagged by our researchers.

The brindle coat is sometimes described as being tiger striped. The color streaks are rather irregular and usually darker than the base coat. This unique variation often leads people to mistake a brindle coyote for a domestic dog or even a wolf.

Searching for the pups fathered by #466 was a highlight every year. Typically, at least half of the litter shared his coloration.

Coyote #466's radio collar lasted seven years but wore out right around the time he died, in October 2017. During those years, #466 rarely wandered more than a mile from his territory within the Crabtree Nature Center. Yet, even today, we receive reports of brindle coyotes in the Spring Creek Valley Forest Preserve, less than four miles from #466's home turf.

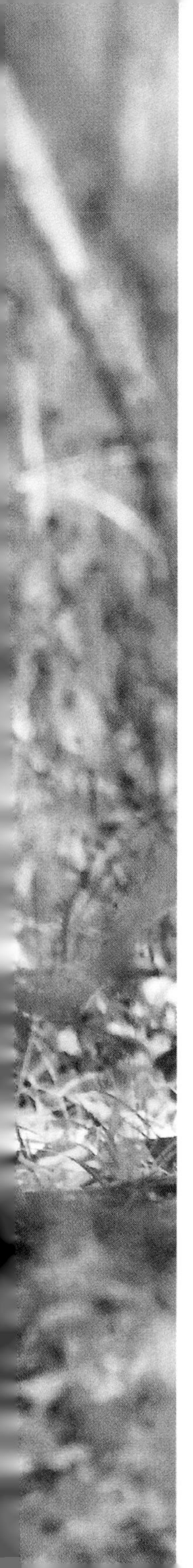

CHAPTER 3

The Indefinable Coyote

Coyotes earn their "ghost dog" nickname. They are enigmas that spend their lives in secret, offering us only passing glimpses as they travel through our landscape, a now-you-see-it, now-you-don't sort of flash seen out of the corner of an eye, a visual trick that makes us question what we saw, if indeed it was anything at all.

Opposite: The coyote's multicolored coat allows it to camouflage across seasons and in a variety of habitats.

If you are lucky enough to observe a coyote for a little while, you may note their resemblance to German shepherds or collies, complete with pointed ears, a slender muzzle, and a drooping bushy tail. They're usually a grayish brown with reddish tinges behind the ears and around the face, but coloration can vary from a silver-gray to black. The tail usually has a black tip. Their eyes are yellow, rather than brown like many domestic dogs. Most adults weigh between 25 and

Group howls are important for family bonding.

35 pounds—only exceptional individuals weigh more than 40 pounds. The largest we have encountered in the Chicago area—a road-killed male—weighed nearly 50 pounds, a truly gigantic animal!

Their size is yet another secret to their success. They are small enough to survive on rodents, yet big enough to take down larger prey. Any smaller and the coyote would be forced to eat only small animals; any larger, rodents and rabbits would not provide enough food. Wolves and mountain lions must remain in areas where they can hunt and kill big game and other large animals; the coyote, which seemingly can survive on anything, lives almost everywhere.

The coyote's haunting howl is part of the American soundtrack, an iconic tone in old Western movies and, today, an element of the urban cacophony. Often mistaken as a yelp of triumph following a successful hunt, the howl is in fact a component of the complex, highly organized social system that governs coyotes' lives no matter where they live.

Their basic social unit is the pack, a family group that works together to defend their territory from other coyotes. In urban areas, a pack usually consists of an alpha male and female, three or four other adults closely related to the alphas, and the pups of the year. In rural areas, the pack may be limited to the alphas and their pups.

Alphas are a pair of mated animals that together defend a territory. They are monogamous, taking on a new partner only after the death of their original mate. While alphas often lead a small group of other coyotes, they also can be a lone pair. Determining which coyotes are alphas is exceedingly difficult—alpha males have no common physical traits, and the only indication that a female might be an alpha is evidence that she has whelped a litter, because only alphas mate. Only after close observation of the coyotes can we determine their status, and even then, we're sometimes wrong.

The packs we have studied live in relatively small, fairly well-defined areas often outlined by roads and other borders, such as the boundaries of forest preserves or parks. Unlike wolves, coyote packs do not travel or hunt as a cohesive unit, preferring instead to roam alone or in pairs. Because of this, coyotes are rarely seen as a group, leading to the mistaken impression that they don't form packs. And it's impossible to determine whether a lone coyote trotting across the field belongs to a pack or is making its own way in the world.

Some coyotes are indeed solitary—about a quarter of those we capture and study each year are not pack members. They are often young males or females between six months and two years old that have been driven from their original family pack, a practice that limits the potential for inbreeding. Others are older animals that have left the packs because they have lost their mate and are seeking another. These "floaters," as we call them, pop up anytime and anywhere, including downtown Chicago.

To survive, the urban coyote must learn to navigate busy streets and highways.

These solitary animals are the itinerants of the coyote world. Moving over large areas, they sneak through multiple territories, avoiding the resident coyotes as much as possible. Because of their transient existence, floaters use a variety of habitats and are quite skilled at crossing roads. We've found them using retention ponds, baseball diamonds, golf courses, patches of ornamental shrubs within parking lots, and small buffer strips between subdivisions.

As for the howl: The coyote is one of the most vocal species of the Canidae, reflected by its scientific name, *Canis latrans*, Latin for "barking dog." Perhaps more than any physical trait, coyotes are recognized by their sounds. In fact, an urban or suburban resident is far more likely to hear a coyote than to see one.

Coyotes make many sounds, but the most common and recognizable is the "yip-howl"—a short yip or bark followed by a longer note. You may hear a single coyote yip-howling or a group chorus. Alphas usually begin the family chorus, and as they hit the long note, other

Residents, Transients, and Dispersers

Coyote populations are composed of three types:

RESIDENTS

Residents live in well-defined territories from which they attempt to exclude other coyotes. These are family groups with a highly structured hierarchy, made up of the breeding pair (alpha male and female) and their offspring. Residents and their territories form the foundation of the local coyote population. All reproduction in the population comes from alpha pairs in residential groups.

LOCAL FLOATERS OR TRANSIENTS

Local floaters or transients have left their groups and territories to seek out a mate and occupy a territory of their own. They have large home ranges that overlap multiple resident territories and even those of other transients. Their challenge is to stay alive while traversing new or risky landscapes and searching for that vacant territory or potential mate. Coyotes may be transients for days, weeks, or, in some cases, years, but throughout that time they are part of the local population, competing for food and shelter. Because they are hard to track and observe, transients are often overlooked, but they play important roles in the dynamics of the local population. For example, transients quickly move in to take the place of coyotes that are removed through predator control efforts or other means, making it nearly impossible to eradicate a local population.

DISPERSERS

Dispersers have permanently left their previous territory and are moving through the population, seeking mates and territories. Unlike transients, they play a very little role in local dynamics because they are not truly part of the local population. They are opportunistic; if they don't find a partner or territory, they will move through an area quickly and not return. Most often they are here today, gone tomorrow. Dispersers sometimes play an important role in gene flow and the reduction of inbreeding.

pack members join in. It quickly devolves into a seeming cacophony of random, high-pitched yips.

Group howling bouts are often performed at "rendezvous" sites within the pack's territory, helping to bring the group together and maintain social bonds. These sites are important features of the pack's territory. If humans hear a neighboring pack in the same location night after night, it is likely a rendezvous site.

Like wolves, coyotes also use group howling to signal to neighboring packs and solitary floaters that their territory is occupied. Often, the howls provoke neighboring packs to respond with their own group howls. Group howling may also convey group size, and some researchers have speculated that coyotes can "count" the number of individuals in a group howl.

Karen Hallberg, a graduate student at The Ohio State University, tested that notion. She played recordings of various-size groups of captive coyotes in areas where we knew radio-collared resident coyotes spent most of their time, and we monitored the residents' responses. In short, resident alphas quickly approached if the playback involved only a pair of coyotes but stayed away or retreated when they heard recordings of a pack of five coyotes. The takeaway? Pack size matters.

The research also uncovered a mystery. Nothing seems to provoke coyotes into noisy reactions more than sirens from emergency vehicles. Of course, such sirens are a notable acoustic feature of urban areas, so coyotes get and give an earful.

Karen stumbled upon a possible explanation when she made the recordings of captive coyotes. To stimulate the animals to howl, she played a siren. Her analysis revealed that emergency sirens often share the exact frequency as an important part of the coyote's call. Just as a coyote's howl will provoke a response, so will the siren.

Coyote packs vary in the timing and tone of their howling choruses. We suspect that packs have group personalities reflecting those of the alphas. This makes sense, as we know that alpha animals have

Social bonds are an important part of a coyote's life.

individual personalities and that nearly all group howling choruses are initiated by an alpha. We have not tested this yet, but it is possible that the tone of howling choruses may be an indicator of differences in pack personalities.

In urban areas, coyotes shelter and hide in wooded patches and shrubbery during the day. Our research has found that, within the urban matrix, coyotes will avoid residential, commercial, and industrial areas but will use any remaining habitat fragments, such as those found in cemeteries, parks, and golf courses. They travel along railroad tracks and the network of parks, road medians, and backyards that thread through a metro area.

They learn to avoid hazards remarkably well. Coyote #441, which we tracked in Chicago's upscale Lincoln Park neighborhood, would quickly cross streets with no traffic, stop and look for oncoming traffic before crossing busier streets, and stop and wait for the traffic lights

Radiolocations of Coyote #434 show that she largely avoided people by restricting her movements during her first year to a marsh and woodlot surrounded by a subdivision.

to turn before crossing the busiest streets of all, strolling through the headlights of waiting cars and doubtlessly thrilling—or frightening—the motorists inside.

In 2015, Dr. Hance Ellington conducted postdoctoral work into the movement and use of space by the coyotes we collared in the most urban parts of the landscape, near downtown Chicago. Using our highly accurate GPS satellite data, he found that the animals' home range was smallest in suburban settings and largest in Chicago's densely populated downtown. "Urban core" coyotes living in or near the city limits had large home ranges, averaging roughly three square miles. However, these urban core animals could not use sizable parts of their territories because of the large buildings, steel, and concrete in the space, so they moved within their territories in long, linear paths. Thus, they had to travel farther each night, and faster, than the other coyotes. Somehow these coyotes maintained exclusive territories, even though they could not patrol their boundaries like their suburban counterparts. Since a coyote's range is related directly to its ability to

"Problem" coyotes are often animals that have lost their fear of humans because the humans are feeding them.

find food and shelter, this suggested that city coyotes must range farther to find food, even if it is plentiful once they do find it.

On average, coyote packs in the suburbs had a range of less than two square miles, while solitary floaters covered an average of fifteen square miles.

There's always an exception, though. We captured and collared Coyote #740, an adult male in a graveyard on the city's north side. Given his large size—42 pounds—and his maturity—at least five years old—we thought he would roam widely across the city, so we outfitted him with an expensive GPS collar to best record his movements. Yet, over the course of the year of GPS tracking, he never stepped out of the cemetery. He didn't have to. As it turned out, two human neighbors had "adopted" #740, leaving food for him every day. Once they put out the food, the neighbors would honk their car horn, a Pavlovian prompt that eventually would bring the coyote out into the open to eat. Surprisingly, this coyote never lost his fear of humans, which prevented him from ever becoming a nuisance.

Coywolves and Coydogs

Sometimes people mistakenly refer to the larger coyotes found in the North and East as "coywolves," suggesting that they are a direct cross between coyotes and eastern wolves. The term started appearing in the 2000s as more people began to take note of urban coyote populations and is the result of some well-intended misinterpretations.

If you take a DNA sample from any coyote east of the Mississippi River, it's likely to show traces of wolf or dog ancestry. The term *coywolf* led to the misinterpretation that these are the products of recent interbreeding to produce a larger animal than the rest of the coyote population, an animal that is half wolf and half coyote. In fact, the trace of wolf DNA is a residual of the early expansion of coyotes eastward, when there was historic interspecies breeding among rare coyotes and wolves in Canada. That DNA spread through the eastern coyote population. All coyotes, rural and urban, in the eastern half of the US possess this small amount of wolf DNA. *Coywolf* is also misleading in that it ignores the fact that geneticists have also found varying trace amounts of domestic dog DNA in eastern coyotes.

As for "coydogs": coyotes and dogs are biologically capable of producing hybrid litters. Some have even been raised in captivity. But genetic surveys have rarely documented evidence of recent breeding between dogs and coyotes in the wild, and in the West, coyotes and dogs have remained distinct despite the fact that domestic dogs and coyotes have shared the continent for the past nine thousand years.

One would think that coyote-dog hybrids might be more common in urban areas with so many dogs present. Although it is possible, coyote-dog hybrids in the city remain rare because:

- Coyotes are highly seasonal breeders with a narrow window for breeding; dogs can mate year-round.
- Coydog females have a shifted estrus cycle that does not coincide with the coyote period.
- Domestic dog and coydog males do not tend to litters, whereas male coyotes do.
- As hybrids, coydogs may have lower sperm viability than either domestic dogs or coyotes.

His GPS data revealed that he never left the cemetery, which meant he had the smallest resident territory of 0.27 square miles. I'm unaware of any other long-term territories that miniscule—which is ironic because he was one of the heaviest and tallest coyotes we have ever captured! His case was an example of the wide variety of lifestyles coyotes lead across Chicago. He created and maintained his own mini-territory for years, receiving free handouts and eschewing potential mates. He was a striking physical specimen and an exception to the rule that feeding causes problems.

People often believe that urban coyotes primarily eat garbage and pets, but this is far from the truth. Although coyotes are predators, they are also opportunistic feeders and shift their diets to take advantage of the most available prey.

Early in our project, we analyzed coyotes' scat to study their diets. This involved collecting feces, or scat, and poking through them to identify the hair, bones, and feathers of prey and plant seeds that are not digested. Luckily, coyotes prefer to defecate in obvious locations, such as in the middle of trails, sidewalks, and roads, making it

Coyote scat is a key to understanding the animals' diets and how they might change in various settings.

relatively easy to collect their scat. (Defecating in such obvious areas is the only thing coyotes do that helps us study them.)

Graduate student Paul Morey led this work, investigating more than 1,400 scats from different locations within our study area. He found that diet items varied across space and time, reflecting the coyote's omnivorous nature. Simply put, they'll eat almost anything.

Suburban and urban coyotes in Cook County had similar diets to rural coyotes, with small rodents, rabbits, and deer making up most of their food, along with some native fruits such as mulberries. Fortunately, we found very little evidence of cats and none of dogs in their diets. These results are quite similar to other urban studies of coyote diets using the same technique, as well as numerous rural studies.

We also study diets by performing necropsies on study animals that died or on animals killed on Chicago's highways and roads. Even after death, these coyotes still provide a wealth of knowledge about their lives by showing us the most accurate picture of what they ate just before death. While we can see only the animal's most recent meal this way, when there is a large enough sample size, a picture begins to emerge for the population. Research technicians had the unenviable task of sorting through 469 stomachs from 2000 to 2022. In most cases, the necropsy results echoed those from the fecal analysis. Rodents are nearly always present in the diet, with a mixture of other items depending on the season, and virtually no pets. We've even found up to twenty-one rodents in a single coyote's stomach! Human foods were relatively rare.

Yet even with all the data we collected from scat and stomach content analysis, we were concerned that these traditional techniques did not reliably detect the presence of human foods in the coyotes' diet and that neither method provides much information on the variation in diet among individual coyotes. So, in 2012, we collaborated with Dr. Seth Newsome at the University of New Mexico to use a relatively new technique called stable isotope analysis.

Stable isotope analysis uses a series of laboratory techniques to measure the isotopic compounds of an element, such as nitrogen or carbon, in an animal's tissue. Each prey animal and plant has its own isotopic ratios for elements that are processed in the coyote's tissue and can be detected in their bone, muscle, or hair.

To begin this process, we collected a whisker from each coyote we captured and sectioned it into fifteen to twenty segments. Each segment was analyzed separately so we could determine whether the animal's diet changed during the growth of the whisker, which for coyotes is likely four to five months. This approach allows us to understand how a coyote's diet might vary with the seasons. We also can cross-reference the results to an individual coyote's tracking data, helping us understand whether a coyote's diet is specific to its environment (such as whether a coyote in an ultra-urban setting eats more human-related food).

Luckily, we found that processed human foods have a carbon isotope signature that is easily discernible from natural foods in coyote tissue. This meant we were able to clearly identify the presence of human foods and natural prey in the diets of 173 Chicago-area coyotes from 2010 to 2019.

All of them had at least a trace of human food in their diets, but 70 percent had diets comprised mostly of natural foods such as rodents or plants. What was remarkable was the tremendous variation among individual diets regarding how much of certain natural foods were consumed (e.g., rabbits, voles, etc.), how much scavenging they did (e.g., deer), and how much human food they consumed. As we suspected, we also observed a higher amount of human food consumption in some coyotes than we had been able to detect through scat analysis. Indeed, 12 percent of the coyotes had diets very much like yours or mine.

Coyotes always surprise us, and they did so again in this study. We assumed that coyotes living in the downtown area would have to exist

In the lab, coyotes' whiskers are collected to determine their diets.

on human foods. Not so. Their diets were as varied as their counterparts in suburbia.

Ours was the first study to use this technique to investigate urban coyote diets, and it was exciting to use this technology to uncover a hidden part of the coyote's urban lifestyle and, especially, their relationship to us. The most important takeaway was the tremendous individual variation in diets, which indicates that coyotes are using a wide range of strategies for exploiting food resources across the urban landscape. The diet variation is a great example of the advantages of being a flexible, highly opportunistic species, and one of the secrets behind the coyote's success. By understanding their diets, we

A typical entrance to a coyote den.

can adjust and alter wildlife management protocols, among other benefits.

Raising a litter is one of the biggest challenges facing urban coyotes. They roam through their territories and seldom are tied to a specific spot—except when they are rearing pups. This makes them more vulnerable to people and other threats, and it's also one of the two times of year when most human-coyote conflicts occur—the other is during mating season, when coyotes are extremely territorial.

Pup season is the only time coyotes voluntarily use dens. Otherwise, they usually sleep above ground, in the open or hidden by grass, brush, or other items. After a gestation period of sixty-two to sixty-five days, a pregnant female will begin looking for existing dens or will start constructing one herself. Dens may consist of a hollowed-out tree stump, a rock outcrop, or an existing burrow made by raccoons, skunks, or other medium-size carnivores. On industrial sites, they've even denned in concrete culverts.

Coyotes will also build a den by digging a hole. They usually prefer protective cover at the den, such as bushes or trees, and a slope for drainage. Some dens are reused for years.

Mothers may move their young from den to den to keep them protected. Some coyotes select secluded areas for their dens, whereas others in more urbanized areas have fewer choices and may establish dens near buildings or roads or in parking lots. We've even located a den in the parking lot of one of Chicago's major sports stadiums!

Once their pups are born, the alphas display strong parenting instincts, perhaps a byproduct of their dedicated monogamy. Their teamwork allows the alpha female to give birth to large litters; we have recorded as many as eleven in a litter from one set of parents. By contrast, solitary species in which the female raises the young alone have smaller litters. The strong parenting instinct among alpha coyotes has also led to some cases of adoption, or double litters in a den. When sneaking into the dens to mark young, we documented three such cases, though we don't know whether the parents rearing the young

Even a hollow log in a residential neighborhood can be home to a family of coyotes.

noticed the difference. Indeed, we know little about this fascinating behavior except that it happens.

We also have observed cases in which the female is lost and the male raises the young alone or with a subordinate. An extreme example is the pair of female #572 and male #575, who resided in Busse Woods Forest Preserve near O'Hare International Airport. In 2014, we located and microchipped each of their seven pups, only to find #572 three days later, killed by a car. It was depressing to think that the cute bundle of pups we saw a few days earlier was now motherless. We crossed our fingers and hoped for the best. The pups were five weeks old, close to weaning. The father stayed in the territory, so it was possible he was trying to care for them. The pups were not radio-collared, so their fate was unknown. However, years later, we captured two of the pups as adults: a daughter, #827, in the same preserve on January 18, 2017; and #824, a male, was shot at O'Hare on February 28, 2018, when he was nearly four years old. We don't know the fates of the other littermates, but these two pups are evidence of successful rearing by a father after the loss of the mother.

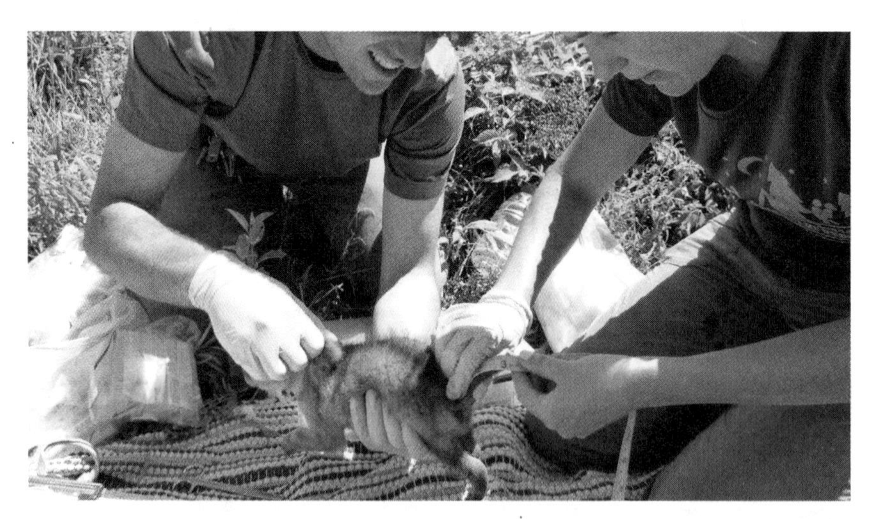

Tagging coyotes as pups helps us understand family relationships and identify them if they are recaptured as adolescents or adults.

In more common situations, pups stay in the den for about six weeks and then begin traveling short distances with the adults. By the end of summer, pups are spending some time away from parents and attempting to hunt on their own or with siblings.

This often leads to what I refer to as "soap opera" moments, when events dramatically change the family dynamics. We can't see the drama unfold. We can only infer what is happening based on tracking locations and other data. Even though we can't see it and don't fully understand it, it is high drama to the coyotes.

At some point, all young coyotes face two important life decisions that determine the trajectory of their lives: 1) when or if to leave the family and the security of their natal territory, and 2) selecting their life partner.

Most coyotes permanently leave their parents and their territory. Whether this is a voluntary decision or one forced by the parents is one of those wonderful mysteries that allow us to imagine the drama behind it. And yes, it is tempting to anthropomorphize—most of us, as parents or children, can relate to the challenges of children leaving the home.

There appears to be no universal answer, other than the fact that nearly all coyotes, male and female, leave once they become sexually active. Many leave during their first fall and winter, and others leave a few months later during their first mating season, in February or March.

Some young coyotes disperse suddenly, apparently without having explored their surroundings, while others seem to test the waters, leaving and coming back until they finally take the leap and leave for good. These exploratory movements may be only as far as a neighboring territory, or they may cover eight to ten miles, only to suddenly reverse course and travel all the way back home.

Why is the choice to leave significant? At the population level, dispersal prevents inbreeding and competition for resources, so it is

Territory Abdication

The best example of abdication is Coyote #1; her mate, #115 (aka Melonhead); and their son, Coyote #227. During their early years together, Coyote #1 and Melonhead maintained a territory that had at its core an area containing massive broadcast towers.

This area was perfect for coyotes—fenced off to the public and maintained as an open grassland to prevent interference with the towers. The alpha pair had dens they used to raise litters near the towers, and it was the pack's focal area from 2001 through 2005. In 2006, the pair began using the area less, and by 2007 stopped using it entirely. They had restricted their movements to the western part of their territory.

At the time, we could not understand why a pair of coyotes that had occupied and defended this ideal area suddenly shifted toward a more urbanized part of their territory. The answer came in 2008, when we captured and radio-collared a handsome adult male under the towers, Coyote #227. He was their son, born in 2005. It quickly became apparent that he was the alpha male of the pack now occupying this prime real estate. His parents, Coyote #1 and Melonhead, gave up the area when #227 matured and continued to maintain an adjacent territory until Coyote #1's death in 2010.

Opposite: Coyote #227. Above: Coyote #1245 traveled one of the farthest recorded distances in our study.

ingrained within the species' genetic makeup. But to the individual, it is a choice between having a secure natal range with little opportunity to reproduce and taking the plunge and engaging in risky travels in hopes of finding a mate. As one might predict, studies across multiple species have shown that survival rates plummet during dispersal. Some examples of dispersal movements from the Chicago area include #679's 70-mile trek north to Milwaukee's Mitchell International Airport, #1245's travel of 105 miles south to Flanagan, Illinois, and #974's dispersal of 114 miles to Milford, Indiana. These are linear distances from their capture locations, so are minimum estimates of the actual distance covered. In each case, the animal was shot by a hunter or at the airport by control officers.

An exceptional few young coyotes, by either luck or design, never leave but take over the territory from their parents. Some individuals take over only a portion of their natal territory. Some can stay by delaying sexual maturity until one parent dies. In these cases, the surviving

A group of coyote pups peek out from their den.

parent is the one that moves on. One can imagine the Shakespearean drama when that takes place.

In the rare instances when an offspring avoids dispersal by taking part of the natal range from its parents, the parents may shift their territory. Chicago's coyotes in those situations more typically split the territory, with the parents and offspring living in smaller portions of the natal range. This is one way that coyote densities can increase. At the beginning of our study, one coyote territory encompassed most of the Poplar Creek Forest Preserve, one of our main study sites. Through subsequent splitting, that original territory eventually became five smaller territories.

A most fascinating situation arises in the rare cases of abdication, in which an offspring takes over prime real estate in their parents' territory while the parents shift or reduce their territory. We have no idea whether the parents abdicate willingly or only after conflict with their sexually mature offspring. If it is done voluntarily, that would be taking parental investment in their offspring to another level.

In these cases, parents and their adult offspring occupy adjacent territories. Over time, with high survival, an area may have multiple territories of related coyotes, but they are just as territorial and exclusive as if they were unrelated. We have also recorded examples of offspring inheriting the natal range upon the death of one, or both, of their parents.

Another "soap opera" moment is the loss of a member of the alpha pair. When an alpha dies, we follow the packmates closely because this will likely affect the dynamics of the group and may precipitate dispersal of the subordinates or pups. One of three things happens: 1) the remaining alpha stays and takes on a new, unrelated mate to continue alpha status; 2) the remaining alpha leaves the territory and new, unrelated individuals take over; and 3) the remaining alpha leaves and one of their offspring assumes alpha status with a new mate.

An example of mate loss and inheritance involves the family of #349 and #678. Female coyote #349 was a longtime resident of Poplar Creek Forest Preserve. She was radio-collared in 2008 and lived to 2017. Her mate was #678, and they had four litters together from 2014 to 2017. We microchipped their last litter of five pups on May 5, 2017. Sadly, #349 was killed by a car in July, just as the pups were weaning. One pup, #1110, was hit by a car a week later, and we assumed that the family was breaking up or the other pups were lost.

To our surprise, later that year, we captured the four remaining pups in early December and radio-collared them. Their father had taken care of them, and they were still living in or near their parents' territory. Pup #1108 soon dispersed and was shot the following year forty-eight miles away near Kenosha, Wisconsin. Pups #1109 and #1112 remained in or near the territory but did not live to reach sexual

Vehicle collisions are a leading cause of death for urban coyotes.

maturity, both hit by cars in 2018. The remaining pup, #1111, a male, never dispersed and inherited part of his natal range. He won the lottery and remained as an alpha male until his death from unknown causes on May 3, 2022.

To better understand the makeup of the coyote population, we collect a tooth from each coyote after they die to get precise ages, which are then used to break down the population into sex and age groups. We need this information to understand whether the population is likely growing, declining, or simply maintaining an equilibrium. Each tooth is sectioned and stained to inspect layers that roughly correspond to years, like tree rings.

Applying this aging criteria to more than three hundred individuals reveals that, like virtually all coyote populations, young animals dominate the Cook County population, with an average age of 1.7 years and a maximum age of 14 years. Usually, the population and each age class are divided equally by sex.

Only 20 percent of coyotes reach three years of age and alpha status, and less than 1 percent live past nine years of age. The accuracy of the aging technique is shaky as coyotes grow up, but the oldest coyote we tracked continuously throughout most of its life (who therefore was of known age) was Coyote #1, who died at eleven years. In captivity, coyotes can live thirteen to fifteen years.

Our research indicates that coyotes of all ages in the Chicago area generally have roughly a 60 percent chance of surviving each year, even juveniles in their first year. Still, the survival rates of juvenile coyotes in Cook County are approximately five times higher than the 13 percent survival rate reported for rural juvenile coyotes. Rural Illinois is dominated by row-crop agriculture, and coyotes are hunted year-round. Once agricultural crops are harvested, coyotes are more visible and, therefore, more vulnerable, especially juveniles. Metropolitan areas, on the other hand, provide more year-round protection since there is almost no hunting and no seasonal loss of habitat due to crop harvests.

Since our project began, 469 out of 1,433 tagged/collared coyotes have died. The causes included vehicle collisions (41 percent), mange (14 percent), harvest (9 percent, though many of those deaths were animals that left the area and were taken in rural areas or around other towns), nuisance control (6 percent), and 8 percent to other causes such as accidents, drownings, and being struck by trains and planes. We were unable to ascertain cause of death for 21 percent of the cases. Thus, about 60 percent of the deaths can be attributed to humans (vehicles, harvest, nuisance removal, and other factors), with vehicles being the primary cause of death across the years and for male and female coyotes of all ages. This high mortality by vehicle makes sense because urban coyotes regularly cross so many roads.

Indeed, among all terrestrial wildlife in Cook County and most North American cities, coyotes have the largest home ranges or movement, resulting in a greater exposure to roads than other animals. This need for space is one of the costs, or challenges, of occupying the "top

In the heart of America's largest cities, coyotes find small remnants of habitat where they can forage, take shelter, and breed.

predator" position in the ecosystem. Restricting their activity to the night helps reduce risk, as traffic decreases dramatically during late-night hours. Coyotes also learn traffic patterns within their territories, as well as safe ways to cross larger roads such as the use of over- and underpasses. Otherwise, the death rate would be much, much higher.

The number of coyotes killed as "nuisance" animals is of particular interest as this measurement is one of our project's primary objectives. By measuring the frequency of nuisance removals, we can better understand the risks coyotes represent. Importantly, half of the cases were animals taken from airports, where there is understandably no tolerance for larger wildlife moving across airstrips. Unfortunately, coyotes find it hard to resist the large grassy expanses that are maintained for airports, especially when the alternatives outside the fences involve asphalt, traffic, and people. To a coyote, the grassy fields must look like a noisy utopia—when, in reality, if they cross the fence, it is only a matter of time before they are killed. Sixteen radio-collared coyotes have been killed at O'Hare. All were shot by control officers except for one that was struck by a plane on a runway. I have come to regard airports, especially O'Hare, as "black holes" for coyotes—an irresistible attraction from which they never return alive.

The other half of the "nuisance" mortalities were mostly coyotes that became habituated to people, or in some cases became aggressive toward dogs.

An important aspect of this project, and a justification for long-term monitoring, is determining whether conflict levels between coyotes and humans might change over time, especially as more coyotes are born and raised in proximity to people. Our results so far suggest this is not happening. The percentage of nuisance-related deaths has remained low throughout the study. If we ignore the airport mortalities, only 2 percent of the radio-collared coyotes have died as part of nuisance conflicts during the study's twenty-three years.

One of the most interesting causes of carnivore mortality is almost absent among Chicago's coyotes: intraspecific strife. That's the term scientists use to classify deaths caused by fighting or attacks by members of their own species. This is often the primary cause of death for large territorial carnivores. But in our study, only one mortality was the result of injuries sustained during an attack by other coyotes, and this was while the animal was in one of our traps. None of the deaths of our radio-collared animals or the unmarked coyotes we found after their deaths could be attributed to their own species.

Even with its many benefits, urban life does increase stress for some coyotes.

This stands in stark contrast to wolves. In Yellowstone National Park, where wolves are protected from hunting, intraspecific strife is the primary cause of death. The attacks are mostly territorial conflicts in which packs kill transients or members of neighboring packs fight each other. Territorial killing is also the primary cause of death for mountain lions and other large carnivores.

All canids possess an instinct to remove competing canid species, and Chicago coyotes display this instinct with foxes and, rarely, with dogs. But they do not kill each other. Even as coyote numbers increased, territories became smaller, and more territories bordered each other over the years of the study, the frequency of intraspecific strife remained low.

Few coyotes reach their full potential life spans in the wild, although there are exceptions such as Coyote #1. She died of natural causes (kidney failure, to be precise), despite spending her life in a heavily urbanized area.

Do coyotes in cities have stressful lives? Most people in large cities would predict that they do. But to date, most urban studies of stress on animals have focused on songbirds with mixed results. And birds are different—they can fly over traffic, can use small patches of habitat, and are generally not threatened by people. What about coyotes?

Graduate student Katie Robertson addressed this question by measuring the stress hormone cortisol in the hair of each of our radio-collared coyotes from 2016 to 2017. Cortisol in mammal hair is believed to reflect levels of chronic stress, which can lead to other complications, such as illness or death.

Katie found similar stress levels between males and females, as well as among different ages of coyotes. Yet stress levels differed by social status. Within the pack, subordinates (which includes resident pups and subadults) had lower stress levels than alphas, which may be the result of the alphas' responsibilities to maintain territorial defenses and protect and raise litters. Subordinates, living within their natal territory under the protection of their parents, appear to live a relatively stress-free life. In Yellowstone wolves, living in a natural system, stress is also higher for alpha animals and lowest for resident subadults.

Not surprisingly, Katie found transient coyotes also had elevated stress levels, on a slightly higher level than alphas. This result makes sense, as transients live with uncertainty, moving over large areas and through established territories while simultaneously attempting to avoid resident coyotes, people, dogs, and most important, vehicles.

Beyond social status, urbanization was indeed related to higher stress: coyotes living in more urban areas had higher stress levels than those living in natural areas or in the suburbs. So, city life does appear to introduce stress to some coyotes, especially alphas. Although we found these trends to be significant, we also observed tremendous variation in cortisol among individuals in the same areas and in the same social class. As with other aspects of their lives, individual coyotes differ dramatically in how they react to the environment, behaviorally and physiologically.

CHAPTER 4

Living with Coyotes

Coyotes get a bum rap. They are almost never noticed until they are perceived as a problem. Then the discussion revolves around the problem—either a coyote that seemingly has lost its fear of humans or has decided to dine on a backyard pet or two.

But predators can't distinguish between wild and domestic prey. These conflicts are almost inevitable as humans crowd into metropolitan areas and coyotes enhance their remarkable ability to coexist nearly invisibly, the ultimate undercover operatives. In doing so, coyotes serve important ecological functions, even in metropolitan areas.

Opposite: In many places, it is common to see a coyote in your own neighborhood.

As the top carnivore in most cities, coyotes potentially impact other animals, either directly through predation or indirectly through competition for food, den sites, or refuge. Over the course of this project, we opportunistically expanded our focus to include some select species to explore their relationships with coyotes in this urban system. These species included other medium-size predators that coyotes may affect

The raccoon is another predator that has adapted remarkably well to urban environments, and, somewhat surprisingly, is seldom attacked by coyotes.

either directly by killing them (either for food or to remove a competitor) or indirectly by forcing them to avoid areas used by coyotes.

The most abundant medium-size predator in midwestern cities has a mask and a ringed tail. Raccoons have been shown to be one of the top urban-adaptive mammals in the United States. Highly intelligent and attracted to novel items and human foods, raccoons can become overabundant in metropolitan areas. Much of our knowledge about raccoon lifestyles comes from our research conducted in parallel to our coyote work. Over fifteen years, we monitored 246 radio-collared raccoons that lived in Chicago-area forest preserves alongside our radio-collared coyotes. Coyotes killed only 2 percent of these raccoons, meaning predation was negligible. We also tested for avoidance. Are so few raccoons prey for coyotes because they are skilled at avoiding them? We used coyote urine to see if raccoons would avoid coyote activity, but it turned out that the opposite was true: raccoons were attracted to coyote markings. Basically, coyotes manage their risk and avoid taking on an adult raccoon.

But the mammal that truly rules urban wildlife, or at least moves through the community with impunity, is the striped skunk. Coyotes are smart enough to know what those black-and-white markings represent, and we have observed coyotes in the suburbs avoiding skunks during our nocturnal radio-tracking. Much like in our raccoon research, we monitored seventy-eight radio-collared skunks during a concurrent six-year study, including within those forest preserves with high numbers of coyotes. No skunks died from coyote predation, and our tests indicated that skunks were unaffected by coyote activity. One of our coyotes fitted with a Crittercam inadvertently came face to face with a skunk. The resulting video showed a split-second view of a skunk face poking out of a den, quickly followed by chaotic blurs of sky and ground as the coyote frantically ran away. So, to the disappointment of some, coyotes will not likely reduce the influx of skunks into some neighborhoods.

In contrast to the coyote's ambivalent relationship with skunks and raccoons, coyotes do have a strong negative impact on foxes. The two fox species found in the Chicago area, red and gray foxes, were quite common in the area before the late 1990s. But as coyote numbers increased, both fox species declined. Foxes were already rare by the time we were able to radio-collar a few, and despite our small sample of nine animals, we confirmed that coyote predation was one factor in their decline. We also observed that foxes avoided areas with coyote activity.

Urban coyotes frequently kill red foxes, a competitor for habitat and food in urban and suburban settings.

Coyotes don't usually kill foxes for food, but rather to remove a competitor. This is the rule in the canid world: the big species removes the small species, and so it goes for wolves limiting coyotes and coyotes limiting foxes. The "coyote effect" on foxes, while certainly a reality, may sometimes be overestimated because other factors such as disease also affect urban foxes. In some cities, foxes have been able to hang on by living as close to people as possible, essentially using humans as shields from coyotes. Much more work is needed to fully understand these dynamics.

Urban coyotes also have an interesting relationship with outdoor cats, an ever-present and challenging species to manage. To uncover the potential impacts of coyotes on cats, I monitored unowned, free-ranging cats and created experimental feral cat shelters in areas of Chicagoland where we also had radio-collared coyotes. We followed humane protocols, and all cats were vaccinated;

A radio-collared domestic cat approaches a colony shelter. Outdoor cats mostly avoided areas used by coyotes.

provided food, water, and shelter; and radio-collared. These were truly feral cats that were unadoptable and had been living "wild" for some time. We found a strong "coyote effect" on the cats, but it was primarily in the form of avoidance rather than predation. Despite the cat shelters we established in areas of high coyote densities, coyotes killed only 7 percent of the 127 radio-collared cats. This surprisingly low predation rate occurred because nearly all cats strongly avoided the green spaces and natural habitat fragments preferred by coyotes and instead restricted their movements to neighborhoods and yards.

Coyote predation of cats is perhaps one of the most controversial aspects of coyote urbanization and often pits sections of the public against each other. Coyotes may kill cats for food or to remove them as potential competitors. Not surprisingly, people who own cats or are otherwise interested in their well-being often consider coyotes to be

Coyote #434: A Human-Made Nuisance

Coyote #434 is a good example of how human behaviors, such as feeding wildlife, can create nuisance coyotes. A young female, she was captured on February 18, 2010, in a marsh surrounded by a subdivision. She was still with her natal pack, which lived in the marsh and a nearby woodlot.

She remained in the marsh and woodlot, carefully avoiding all residential properties and using a power line easement as a corridor through which to move. This continued until August 2010, when she began traveling outside the pack's territory. She was a solitary floater during September and October 2010.

On November 3, 2010, the Illinois Department of Natural Resources asked us to follow up with a woman complaining about a collared coyote appearing in her backyard each morning. The animal was not threatening but was attracted to the squirrels beneath some bird feeders. She also was eating food put out for deer, cats, and other animals—even crusts of bread. It was #434.

The access to food left out for other animals led #434 to change her behavior—instead of moving mostly at night and carefully avoiding residential areas, she now was entering yards during the day. A "nuisance" coyote had been created by humans.

Coyote #434 began showing up in residential areas when food was put out.

an unnecessary threat. Yet our results suggest that, at least for feral cats, coyote predation is not as common as predicted.

Equally important, our findings suggest there could be a trickle-down "coyote effect" that many would consider positive. Studies in Californian urban areas showed that when coyotes limited cat populations, songbirds enjoyed better nesting success. Thus, the coyote's removal of an important smaller predator helped birds and perhaps other species. Essentially, coyotes served as ecological buffers for green spaces, limiting use by outdoor cats and potentially benefiting a variety of birds and small mammals that traditionally are cat prey. Other studies have found increased diversity of native wildlife species in urban landscapes where coyotes are present, largely due to their impact on outdoor cats.

Our diet studies showed that most Chicago coyotes are still focused on prey species despite the abundance of human food available

A coyote "mousing."

Coyotes may help decrease the rodent population in cities, but more research is needed.

to them. Rodents, such as mice and voles, make up the bulk of the coyote diet in both urban and rural areas. Even some coyotes residing near the downtown area consumed rodents, usually along easements bordering major roads or alongside train tracks. Although the impact of coyotes on rodent species hasn't been adequately measured in urban systems, a few studies in rural areas have shown that removing coyotes results in a dramatic increase in rodent abundance and a decrease in rodent diversity, meaning that only a few species increase, to the exclusion of other rodent species. Rodent increases have been observed in certain urban areas, such as golf courses, following coyote removal programs.

There is also the possibility that coyotes help to control woodchucks, also known as groundhogs. Many areas, including cemeteries and golf courses, have reported declines in woodchuck abundance once coyotes appeared. Whether that is a good thing likely depends on your perspective. A golf course groundskeeper, for example, would probably be thankful to be rid of the tunneling groundhogs.

For years, we could not document an instance of an urban coyote eating that other highly adaptable urban creature: the rat. It is important to note that the city's rats, known as brown rats, are not native to North America and are found only in the most heavily urbanized parts of the city. Chicago's rat population is so abundant that the city is sometimes considered the nation's "rat capital." In the early years of this study, I was often asked whether coyotes would help rid the city of its rats, and my answer usually provoked surprise and some disappointment. We didn't know exactly why rats were absent from their diet, but it probably was a good thing for Chicago coyotes. The city had an aggressive rat-poisoning program, and by passing on rats, coyotes avoided yet another human-made threat to their existence. Studies in other cities have also reported limited consumption of rats by coyotes, including a recent study in New York City, also known for its abundant rat population.

Yet even after years of study and research, coyotes still surprise us. In 2020, we captured an adult coyote in Chicago and watched as it regurgitated an entire rat. This case may be an outlier, or an indicator that some coyotes may be shifting their diet.

If I had to speculate, I believe the reason coyotes rarely consume rats is availability. Although Chicago has an abundant population, the brown rat is completely dependent on human food and buildings, sewers, alleyways, and so on. The rats live in and under buildings, whereas coyotes forage in green spaces and avoid humans as much as possible.

At different stages of our study, we were able to expand our objectives to explore the relationship between coyotes and two specific but wildly different urban prey species: white-tailed deer and Canada geese. In both cases, coyote predation has helped slow the population growth of the prey species at the local level.

Deer are often overabundant in urban areas and difficult to

A fawn and mother are radio-collared as part of our study of coyote predation of deer in Chicago. The fawn's collar is modified to expand as it grows to adulthood.

manage. Although coyotes rarely take adult deer, they are primary predators of fawns. Between 2013 and 2018, we monitored fawn survival in our study sites with radio-collared coyotes. This was incredibly time-consuming, difficult work requiring us to locate a fawn within a few hours of birth and then fit it with a modified light, expandable radio collar within a few minutes so we didn't attract the attention of predators. Once fawns were radio-collared, we had to check on them at regular intervals day and night to recover their remains if they died. We found that coyote predation was the primary cause of fawn death each year and that, in most years, coyotes killed more than 50 percent of fawns in our study area. In some years, it was as high as 80 percent.

Nearly all coyote predation of fawns was in the first four weeks following birth. If fawns can survive that first month, their survival rate jumps to 90 percent. Once they become adults, they have little to fear from coyotes: predation by them is minimal.

Is there any animal that pulls at our heartstrings as much as a newborn fawn? Fawns are as cute as any other young animal, and despite our efforts to maintain our professionalism, it became depressing to document their death to coyote predation. Likewise, public relations for coyotes take a hit when the public finds part of a fawn on their lawn. As difficult as it can be to experience, this service by coyotes is important and helpful to humans. The predation rate of more than 50 percent of fawns is sufficient to limit deer population growth at the local level—and limiting the deer population has direct benefits for people. The most dangerous wildlife species to urban residents and their property is not a predator but rather deer, due to their collisions with automobiles. It can be hard to conceptualize, but that adorable fawn represents a potential danger much greater than coyotes or other urban animals. Each year, especially in large urban centers, tens of thousands of accidents occur with deer, causing injuries to people and occasionally fatalities. For example, the Illinois Department of Transportation reported that during 2021 there were more than 14,500 auto-deer accidents across Illinois, resulting in 584 injuries and two deaths. By comparison, there have only been two recorded human deaths from coyote attacks over the past fifty years across the United States and Canada.

There is a particular irony here. Through a largely unnoticed process, a predator that the public associates with risk in fact helps reduce the much more substantial risk posed by a species that is not perceived as a risk. When I talk to people about urban coyotes, I often point out that it is very likely that coyotes regularly save human lives by reducing deer populations. So, too, do the coyotes reduce the chance of deer devastating gardens.

Coyotes' predation of Canada geese also benefits humans. Our research into the relationship between Chicago coyotes and Canada

A trail cam captured a coyote stealing a goose egg at night.

geese generated one of the more fascinating and surprising findings of our research. Canada geese have adapted to urban landscapes much like deer and, at times, become overabundant and a nuisance. Like deer, geese can also be a challenge to manage in urban areas. Until ours, there had been no studies linking coyotes to urban Canada goose populations.

By coincidence, waterfowl researchers conducted a large-scale study of the goose population across the Chicago metropolitan area concurrent with the early stages of our coyote study. Theirs was a massive effort, tracking geese of all ages and monitoring nests across portions of six counties. They determined that the population was growing at a slower rate than expected, and that nest predation was a major factor. However, they didn't design the study to identify predators.

By chance, the biologist leading the goose study discussed the surprising results with me. I recognized that my tracking information suggested coyotes might play a role, but to what extent was a mystery. Areas that had been safe to nest in during prior years were now experiencing high predation rates. Most striking, geese love to nest on islands in urban lakes, with water serving as natural protection from predators, but during their study, geese across the metro area, including on islands, were struggling to reproduce. Could coyotes be responsible? Could they actually swim out to islands?

Graduate student Justin Brown accepted the challenge to try to identify what species were responsible for the nest predation, a difficult task that required creativity and technology, including our own versions of infrared spy cameras and Plasticine eggs. Perhaps most impressively, Justin also marked each egg in the nests, all the while parent geese attacked him. He found that roughly half of all nests were raided by predators, and that in 78 percent of cases that predator was a coyote. Raccoons were the only other nest predator, but importantly, much of the raccoon predation took place only after the nesting geese were displaced by coyotes; thus, the "coyote effect" was both direct and indirect. Justin also discovered coyotes caching the eggs, which allowed them to continue consuming eggs for some time after nesting season. In a few cases, coyotes also killed geese tending to nests. Using these data, we confirmed that the "coyote effect" was largely responsible for significantly slowing the growth rate of the region's goose population.

Geese are formidable nest defenders, and our cameras revealed that they could easily ward off any predator except coyotes. Thus, coyotes are serving as a biocontrol for urban geese. As with deer, coyotes do not take enough adult geese to reduce the population, but they can slow the population increase through egg predation. Eggs are the perfect fast food for coyotes—loaded with protein and fat and packed in a handy container. Removing eggs is so effective in controlling Canada goose populations that human wildlife managers do it too—a process known as *addling*.

An important lesson from our study of the coyote-goose relationship is that this predator-prey conflict takes place each year across the region and goes unnoticed among nine million people. It wasn't until we were able to use technology, such as infrared cameras, that we could observe it ourselves. You may have seen some parks or commercial properties install fake coyote silhouettes to deter geese. Our study inspired this tactic.

Although these positive aspects are important, they are only the beginning of our understanding of the ways in which coyotes positively affect the urban ecosystem. The investigations mentioned here were incredibly challenging, took years of effort, and required the best technologies. We are only scratching the surface of the full picture of what coyotes bring to the urban ecosystem and the roles they play across North America. Unfortunately, most funding for research on coyotes and other mammalian predators has been focused specifically on conflicts and ways to control or limit their populations. This leaves us with an incomplete understanding of how coyotes function ecologically and how we benefit from them.

Despite these known and unknown benefits, coyotes also bring challenges. A major finding from our two-decades-plus of research is the extent to which coyotes and people are living together; more coyotes have been observed using developed areas than expected. People are unknowingly in close contact with coyotes each day, and in most cases, the coyotes are ghostly specters, floating unnoticed past unsuspecting humans, much as they did on the plains. But coyotes are watching and learning from us; we influence their behavior, and our actions will determine their future.

Citizens and some local administrations complain that coyotes don't belong in cities, or that there are too many. In other words, even though the risk of an attack is low, why should humans put up with them at all?

A replica of a coyote may help keep geese away.

My answer, of course, is that coyotes themselves determined that they belong in cities. Coyotes immigrated into cities on their own, usually in the face of extreme persecution. They are part of the urban landscape.

We have estimated the Cook County coyote population to range from two thousand to four thousand animals. That may seem large, but it isn't. People sometimes complain that something needs to be done to control their numbers. In response, we offer some perspective. The highest, most outrageous densities we have recorded for coyotes in some forest preserves so far are around sixteen per square mile. That is much, much, much higher than other published studies for coyotes in rural or urban areas.

But what does that number really mean? These high densities occur in some of the Cook County forest preserves, not in residential

Avoiding Conflicts

Fortunately, coyotes only rarely become aggressive toward humans and pets, but there are several things that urbanites can do to prevent the conflict in the first place.

DON'T FEED COYOTES.

The most effective way to prevent coyote attacks in your neighborhood is simply to not feed wildlife. Coyotes fed in residential neighborhoods can lose their fear of people and may eventually test humans (and pets) as possible prey.

Avoid leaving out pet food or garbage or owning large bird feeders. Though coyotes don't usually eat birdseed, the feeders often attract rodents that in turn attract coyotes.

Tie down your garbage-can lids and avoid putting out garbage for pickup at night. Composting bins should be enclosed and vegetable gardens fenced. Harvest ripe fruit immediately and pick up fallen fruit from the ground.

DON'T LET PETS RUN LOOSE.

Wherever you are, coyotes probably live nearby—so don't let your pets run loose. Keep dogs on leashes and don't leave small pets unattended outside. Electronic and traditional yard fences may keep your pets contained but they do not always keep other animals away.

Free-ranging cats may also attract coyotes. Keep domestic cats indoors and have feral cats spayed or neutered to control their population.

In recent years, many suburban and exurban families have built chicken coops on their properties only to lose their hens to predators. Remember that coyotes are attracted to any available food source, and unattended, free-ranging chickens are surely a temptation.

CONSIDER REPELLENTS OR FENCING.

Some repellents may keep coyotes out of small areas such as yards, although these have not been tested thoroughly. Repellents may involve remotely activated lights or sound-making devices. Sprays have not proved effective.

Normal fencing may not be enough to keep coyotes out of a yard. Consider a fence that is more than six feet in height with a rolling bar across the top to prevent the coyotes from pulling themselves up and over the barrier.

Shut off crawl spaces as coyotes may use them for dens.

DON'T CREATE CONFLICT.

If a coyote is acting as a coyote should by avoiding humans and pets, don't haze or otherwise aggravate the animal. Leave it alone!

REPORT AGGRESSIVE, FEARLESS COYOTES IMMEDIATELY.

Coyotes seen in the day may have grown accustomed to humans and therefore may be more likely to attack. Yell, wave your arms, and/or throw something at the coyote. If you're wearing a coat, spread it out behind you to make yourself appear larger, and calmly walk away. Do not run.

When a coyote fails to exhibit fear of humans or acts aggressively, the animal should be reported as soon as possible. Signs of aggression include agitated, unprovoked barking, raised hackles, snarling, growling, and lunging.

If you are having a conflict with a coyote, you may need to contact your animal control or police department to learn about their protocols for handling coyote issues. Each municipality and agency may respond differently.

In many instances, removal of an undangerous coyote (i.e., one that is simply present but not causing harm) will be the property owner's responsibility. In this case, you must hire a licensed wildlife trapper. Wildlife handling should always be conducted by a professional.

areas, where their numbers are much lower. And how does this compare to other wildlife species in Cook County?

Here are the "worst case" density estimates for a variety of wildlife species in Cook County, coming either directly from our own research or from independent reports:

- Coyotes: 5–15 per mi^2
- Raccoons: 180–260 per mi^2
- White-tailed deer: 50–155 per mi^2
- Canada geese: 60–140 per mi^2
- Feral cats: 25–40 per mi^2

Coyotes are far less common than other wildlife species, even at their highest densities. Coyotes will always be scarcer than these other species because they don't overpopulate an area. Their territorial space system limits population size, and so does their mating system, in which only one pair mates per territory. I think this gets overlooked when people fixate on a population size of a few thousand across a county of 1,635 square miles.

The good news is that the public has some control over how many coyotes occur in a local area. Coyotes will not live in or use an area where there is no food, and food determines how many live in any given spot. If people put out food for coyotes, then more coyotes than usual will stay and use that area. Food provisioning either causes an actual increase in coyote numbers by attracting coyotes and keeping them there or causes a perceived increase by changing coyote behavior (in these cases, the local population does not change, but people think it does because the animals change their behavior in response to the food). We have recorded many cases of this, and in fact people feeding wild coyotes is a problem that never goes away.

People can also reduce the local population through removal, although the reduction is likely to be temporary—and how temporary

A vest can help protect dogs from coyotes.

it will be is influenced by many factors. We have also documented the effects of removal in many cases, and our results have influenced the practices and beliefs of multiple animal control agencies and city administrators.

The take-home point is that citizens should feel they have some control over the coyote population that most affects them (their local one), even if they have very little control over the larger population. The overall population number is of limited importance.

One of the worst and most publicized conflicts between humans and coyotes is the occasional attack on a domestic pet. It's not at all unusual to read or hear accounts of missing pets with the underlying assumption that coyotes killed and ate the animal. In fact, it is extremely rare for coyotes to consume a dog they kill—so why do they sometimes attack dogs?

As mentioned earlier, within the Canidae, there is an instinctual rule that all species follow: the bigger dog removes the smaller dog. Under this hierarchy, the wolf kills or excludes the smaller coyote, and the coyote kills or excludes the smaller fox. In the West, where wolf populations have recovered, wolves are a primary cause of death among coyotes and have reduced coyote densities from pre-wolf numbers. Similarly, coyotes are a threat to the diminutive kit and swift foxes in the West and the red and gray foxes in the Midwest.

The instinct to remove potential competitors is strongly ingrained in these canids' DNA. Thus, when an urban coyote initiates an unprovoked attack on a dog, it is not predation per se. Nor are they trying to kill for food. Rather, they are following their instinct to remove competitors and potential threats to their offspring.

This rule of hierarchy can be complex, and larger species do not always attack the smaller ones. In some cases, smaller species can even benefit from the larger species, such as coyotes scavenging wolf kills in Yellowstone National Park. But the threat of larger animals always exists, and smaller species must be wary.

Given this ingrained behavior, it is remarkable that coyote attacks do not happen more frequently. Indeed, our observations of radio-collared coyotes active under the cover of night have revealed that, in most cases, coyotes make every effort to avoid dogs, even the very small, loud ones. This is the norm across urban landscapes, where, to avoid conflict, coyotes somehow control their instinct. When exceptions occur and a coyote attacks a dog, doing so is the coyote's natural behavior, following its instinct to remove a competitor.

Indeed, it is remarkable that coyote attacks on dogs are so rare. Every day and night, people walk their dogs past hidden coyotes. At night, coyotes pass by yards containing dogs that often bark at them as they go by. Yet in most cases, the coyotes resist their instincts and, instead, avoid us.

Early in our study, we searched newspaper databases for reports on pet attacks in the Chicago area. Anecdotally, we found that reported

Coyote #736: "Pug"

Sometimes good parenting results in conflict. Coyote #736 was a healthy young male trapped and radio-collared on June 2, 2013, on Chicago's Southwest Side. His tail had a unique deformity—it was short with a twist that resembled a pug's tail, leading us to nickname him "Pug." He was apparently a mate to an unmarked female, and they had a litter when we caught him.

Pug traversed some of the most urban, rugged streets of Chicago, primarily at night, and the expanse of land he covered was remarkable. But we studied him for only eight days.

Pug frequented an area where Canada geese had become a nuisance. Dogs were brought in to chase off the geese, but they also disrupted the area's coyotes. Given the time of year, it's likely that the coyotes were defending young pups.

The property owner decided that the coyotes had to go, despite their natural role as a goose predator. Pug and other coyotes were shot, even though there had been no reports of the coyotes harming the dogs or people.

We tracked him for only a short time, but Pug provided valuable information for our project. He also was featured in National Geographic Wild's show *Urban Jungle*. If you ever catch the episode "Downtown," you will see him and his iconic tail.

incidences grew through the 1990s and early 2000s—the period when coyotes were beginning to infiltrate Chicago.

Our review had the following points: Coyote attacks can happen at any time of the year, but they escalate during February and April, which coincides with mating and litter-rearing seasons. We found cases of coyotes attacking almost thirty breeds of dogs. Smaller breeds such as Yorkshire terriers, shih tzus, and Jack Russell terriers were the most frequent targets, but coyotes also took on breeds twice their size, such as Labrador and golden retrievers. Perhaps because of this size differential, the larger breeds were usually attacked by two or more coyotes, often alpha pairs. As might be expected, attacks on smaller dog breeds were more often fatal and usually involved a single coyote. Attacks took place while dogs were in their backyard (both alone and in their owner's presence) and while being walked in parks.

But it is a myth that a single coyote often lures dogs into ambushes to be killed by its entire pack. More likely, only a single coyote is spotted but the rest of the pack is nearby. When the coyote runs away from the dog, it heads for the safety of the group. It almost surely isn't a planned maneuver, just coincidence.

Opposite: Coyotes at the side of the road in Death Valley, California.

To prevent attacks, dog owners should be cautious about leaving dogs—especially smaller breeds—alone outdoors. Fences can help to keep coyotes out of your yard, though they have been known to jump them. The best fences for keeping out coyotes are at least six feet tall and have a roll bar on top.

There are many good reasons to keep dogs leashed when walking in parks: to keep other animals and dogs safe from them, and to keep dogs safe from traffic, each other, and possible predators. Dog owners who walk in parks known to be frequented by coyotes can take the extra precaution of carrying a walking stick, noisemaker, or spray to fend off a possible attack—although an attack by another domestic dog is far more likely than one by a coyote. Remember that coyotes' behavior tends to change around breeding and litter-raising season and, above all, make sure that no one in your neighborhood is attracting coyotes by leaving food outdoors.

Although coyotes do attack and kill domestic cats, such attacks are often difficult to substantiate and challenging to prevent. The best way to protect your cat—from a coyote attack, or from the far more likely chance that it could be killed in traffic—is to always keep it indoors. If you must let your cat outside, do not leave food for it or other animals outside. That food can attract coyotes, which may then attack your pet.

STOP

The most extreme form of conflict between humans and coyotes concerns attacks on people. These are very rare, and many times coyotes are blamed for attacks by domestic dogs. One such example came in 2012, when newspapers and other media outlets reported that a man in suburban Aurora, Illinois, had been bitten by a coyote.

Subsequent DNA testing of the animal's saliva confirmed that a domestic dog was responsible. Dogs historically are to blame for all canid bites in the Chicago area; the central issue in this case was a human problem stemming from a lack of responsibility among pet owners.

A year later, another report came that a boy on Chicago's West Side had been bitten by a coyote. Once again, the culprit was a dog. Still, several coyotes were removed as a precaution, and several residents confessed that they had been feeding the coyotes—a practice that can cause the animals to lose their fear of people and lead to attacks.

As part of our research, we analyzed coyote attacks throughout the United States and Canada between 1985 and 2006. We collected accounts of 142 incidents, resulting in 159 human victims over a wide geographic area, including fourteen states in the US and four Canadian provinces. Most attacks occurred in the western US, with almost half occurring in California and another large portion occurring in Arizona. There were no reported differences in the frequency of attacks between women and men or between adults and children. However, children were the victims of the most serious attacks.

Attacks generally fell into five categories: 1) defensive—the coyote felt threatened and was defending itself, its pups, or a den; 2) rabid—the coyote was captured, tested, and diagnosed with rabies; 3) pet-related—the coyote attacked a person who was walking a pet, trying to save a pet from a coyote attack, or just near a domestic pet at the time of the attack; 4) investigative—the coyote bit a sleeping or resting person, testing it as a possible prey source; and 5) predatory—the coyote directly and aggressively pursued and bit the victim.

We classified 37 percent of attacks as predatory, 22 percent as investigative, 7 percent as rabid, 6 percent as pet-related, and 4 percent as defensive. The remaining 24 percent could not be classified due to a lack of detail. Predatory attacks caused the most serious injuries. Most victims were doing some sort of recreational activity (e.g., camping, walking, or biking). Others were relaxing outside their homes, sitting on porches, grilling out, or sleeping outside. Most attacks were near the victim's home or in parks.

Slightly more attacks occurred during the January through April breeding season than in other months. We did not find a meaningful difference in the number of attacks occurring during daytime versus nighttime hours.

Most victims were attacked by seemingly healthy coyotes; only fifteen victims in our study were bitten by rabid coyotes. Most did not suffer serious injuries and were able to get away or scare off the coyote by yelling or throwing objects at it.

Only two fatal coyote attacks in the US and Canada have been recorded in modern history: in 1981, a three-year-old California girl died of injuries sustained from a coyote attack, and in 2009, a group of eastern coyotes killed a nineteen-year-old woman who was hiking alone in Cape Breton Highlands National Park, Nova Scotia. The three-year-old's case was predictable in hindsight, as the neighborhood had been feeding a group of coyotes that eventually became too comfortable around people. But in 1981, the risk was not fully understood.

I helped to investigate the second attack, the only documented case of an adult fatality and one of the most extensively examined coyote incidents on record. Many people feared that coyotes were changing their behavior to target humans and that this attack was only the beginning of a highly dangerous trend. Fortunately, we found that the Cape Breton system was unique ecologically, in that food and resources were very limited. In metropolitan areas, the opposite is true, and this explains why unprovoked attacks on people are so rare in cities.

A coyote's natural reaction to people is to hide in heavy brush, among rocks, in crevices—anything to avoid humans.

Often, humans are the indirect cause of coyote attacks. In almost one-third of the reported attacks, coyotes were being fed either intentionally or accidentally near the attack site. One victim was bitten while feeding a coyote and another was bitten by a coyote that was being fed by her parents. After speaking to wildlife officials in areas of known coyote attacks on humans, we strongly suspect that wildlife feeding played a role in many other cases.

Our behavior directly affects our relationship with coyotes. Coyotes react to us, and we can foster mutual respect or a lack of respect through cues we send to coyotes. Therefore, urban coyote management begins with public education and untangling facts from myths. People should understand the differences between true threats, harmful interaction, and healthy coexistence.

There is no doubt that the coyote is fascinating, yet some people like seeing and interacting with coyotes so much that they feed them

Caught in the flash of a trail camera, Coyote #466 appears to be the epitome of a "ghost dog."

or even attempt to bait them and train them. Other times people inadvertently feed coyotes by leaving pet food outside and maintaining large bird feeders that attract squirrels and rodents as well as birds. Feeding and interacting with coyotes habituates them to humans in a way that is dangerous for both species.

A coyote that has lost its fear of humans quickly becomes a problem. When a coyote is habituated to humans, it's important to do everything you can to re-instill their fear of us. At times, we have used noisemakers, such as cans filled with rocks, to scare coyotes. Other times we have shot at them with paintball guns. We've also experimented by putting out trail cameras with flashes to scare coyotes from residential yards, but with limited success.

Even normally calm animals can change behavior based on the seasons. A coyote may demonstrate defensive behavior such as barking or, more rarely, standing its ground and not retreating as you

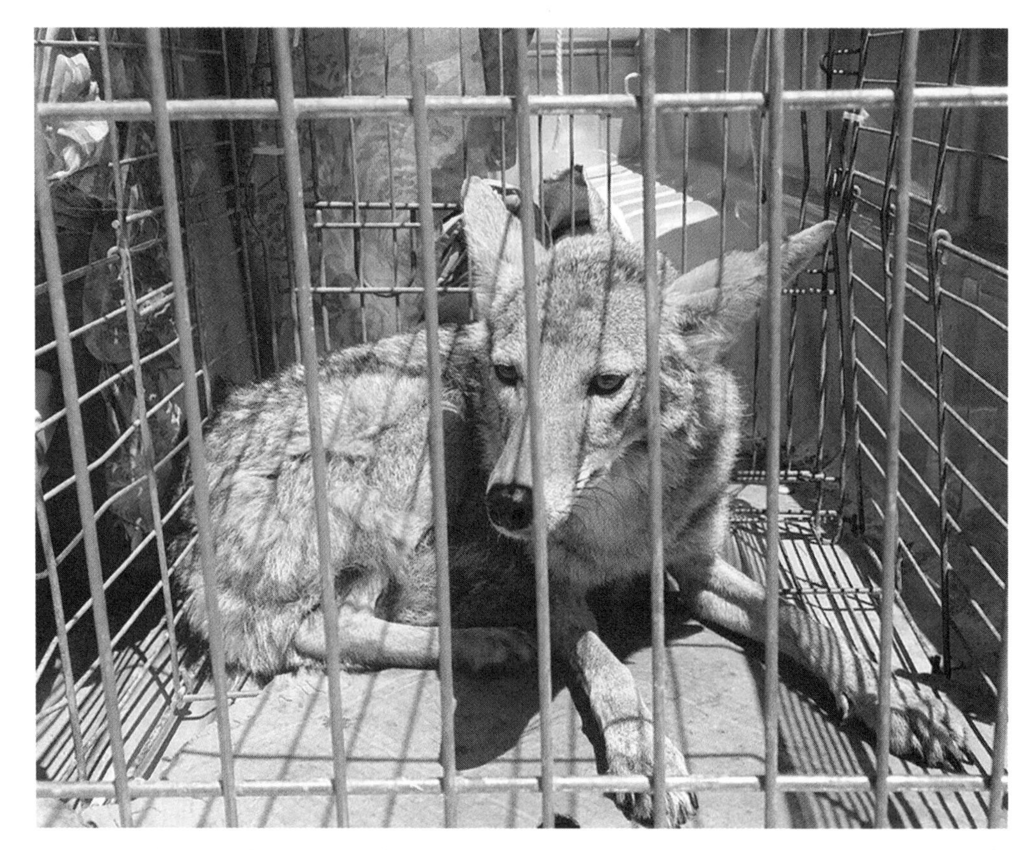

Coyotes can be trapped and relocated, but only by professionals.

approach. If a coyote seems intent on defending a certain area, particularly around the late-spring pupping season, your best bet is to walk somewhere else.

In some cases, coyote habituation is so severe that the coyote can be considered an immediate threat to people, especially children and pets. This is when removal is warranted by trapping or shooting. Neither is easy, and both should be undertaken strictly by professionals, especially in urban areas. Removal efforts are also subject to state and municipal codes.

Targeted removal of one nuisance coyote is far more successful than attempting to wipe out an entire population of coyotes. Because they are not afraid of people, nuisance coyotes are often easier to capture. It is difficult at best to capture all coyotes residing in an area. Even if many coyotes are removed, others move in—and they may be even more used to humans, presenting a potentially bigger threat than the original nuisance animal.

In addition to being ineffective, wholesale removal programs can also be expensive and their traps can occasionally capture pets. For these reasons, coyote removal is best employed only after education efforts or if there is an immediate and obvious safety threat.

Another option is to remove coyotes with live-trapping and then relocate them to a distant site. Though it is more palatable to the public than lethal removal, relocation is rarely effective for any species, especially coyotes.

To measure the efficiency of relocation, we monitored twelve coyotes that were trapped and released in new areas. Even though they were placed in good habitats, none of them remained at their release site. Each traveled in the general direction of where they were captured—but none reached their goal. Cars or hunters killed most of the animals soon after they left the release site. Despite the lack of success, many jurisdictions still relocate coyotes with the understanding that it will likely result in the animal's death.

The fact remains that the best way to avoid coyote-human conflicts is to prevent them by good trash management, keeping pets supervised and on leash, and not providing food to coyotes, intentionally or otherwise.

Despite our preoccupation with coyotes' threats to pets and humans, the reality is that coyotes are under far more threat from us. In its original habitat, the coyote's main threat came from other animal predators—mountain lions, bears, and wolves. But humans have proved to be the coyote's most relentless threat by far. By intent or by accident, we are the coyote's greatest predator.

Federal and state authorities kill tens of thousands of coyotes every

Coyotes #881 and #882: Evicted by Bulldozer

We captured two adult coyotes on the morning of April 8, 2015, on Chicago's West Side. As we tracked them, it soon became clear that they were a mated pair. The female, #881, was between four and six years old, and her mate, #882, was a year or two younger. The female was pregnant, and we expected her to whelp within the next few weeks.

Over the next month, we came to believe that this pair would make their den in a small grove of trees in a lot adjacent to the Chicago River. The site had everything a coyote would want, including easy concealment. Our suspicions were confirmed in mid-May, when our radio trackers showed that Coyote #881 was indeed at the site.

The next day, our researchers drove to the site and let out a collective gasp. Two front-loading bulldozers had leveled the site, leaving only two heaps of debris. The den site was gone, and so was #881. Our radio tracking told us she had moved about two hundred yards away, on the banks of the river. The pups were dead; otherwise, she never would have left the area. Today, her den site is a community garden.

Coyote #881 crosses the street in a residential area.

year, easily the most of any species targeted by government management programs. In 2021, the US Department of Agriculture's Wildlife Services reported killing 64,131 coyotes, compared to 33 black bears, 200 mountain lions, 605 bobcats, 3,014 foxes, and 24,687 beavers. Many were shot by government sharpshooters in helicopters, mostly to protect livestock and big game.

Others are trapped, both as a means of wildlife control and also for their fur. The market for animal pelts varies widely with international demand. Russia and China in particular buy furs for fashion. In North America, one of the most popular winter coats features a hood trimmed with warm coyote fur. The manufacturer announced in 2020 that it planned to stop buying new coyote fur but would not give it up entirely. Instead, it would buy back used coats and recycle the hoods. Other clothing companies continue to use new coyote fur.

Lastly, hundreds of thousands of coyotes are killed annually for sport, by hunters who relish the challenge of luring a wary coyote into gunshot range and by those who are convinced that coyotes will harass and kill the deer they would like to hunt themselves. There are even organized coyote-killing contests that can involve trucks, hounds, and hundreds of participants. In some cases, the dead animals are skinned and their pelts sold; in others, they are treated as vermin and destroyed. In some areas, particularly the West, coyote carcasses are strung up on barbed-wire fences. Niche TV shows focus on coyote hunting, and videos of "kill shots" populate YouTube and other social media platforms. *Field & Stream* magazine published an online article with a headline that seemed to sum up the prevailing attitude: "Are Coyotes Murderous Villains or Worthy Adversaries?"

If we combine the number of nuisance animals removed by local animal control officials and private predator services, the coyotes that are hunted for sport or fur, and the many coyotes killed by cars and other unintentional human interactions, the number of coyotes killed

by humans each year likely exceeds one million—and that is a conservative guess. Up to 70 percent of coyote deaths each year happen on the roads.

Does all this do anything to eliminate coyotes, or even to reduce their long-term populations? The answer seems to be no.

There always will be coyotes that must be lethally removed. Some are indeed threats to livestock or have diseases such as rabies or distemper. Others have lost their fear of humans and should be considered potentially dangerous. In addition, licensed, regulated hunting and trapping is carefully monitored by wildlife officials to ensure that the overall population is not threatened.

Yet even where coyotes can be killed year-round and without limits, the coyote thrives. This seeming contradiction is explained by the coyote's remarkable history and ability to adapt.

Opposite: A coyote observes the city's hustle and bustle.

Nature abhors a vacuum, and so does the coyote. As other predators disappeared, coyotes showed up and filled an ecological niche across the continent. The same thing occurs when coyotes die or are removed from an area. If it is a good habitat, other coyotes will soon be along to claim it. If the habitat offers sufficient food and cover, the alpha pair in a pack will produce larger litters each year, repopulating the landscape.

Among mammals, only the coyote has been able to survive such concentrated and long-term persecution. They can live almost anywhere; they can eat almost anything. Poison them, shoot them, trap them—coyotes come back. It is their superpower.

Bub's
READ HERE
Avid

CONCLUSION

Moving Forward

The phenomenon of the urban coyote is largely a great, unplanned, "natural" experiment initiated by coyotes, with humans acting as involuntary, often reluctant, participants. Clearly, this experiment is changing our relationship with coyotes. Our research over the past two decades, as well as the work of many other scientists, has attempted to document this new ride we are on and to provide keys to understanding how it is happening and what it means for us and for coyotes.

Opposite: Graduate student Ashlyn Halseth microchipping a coyote pup.

When the Urban Coyote Research Project began in 2000, it was one of the few studies of its kind in North America. While our work continues—we are the longest-running coyote study on the continent—others have begun, mirroring the coyote's spread. Some are ongoing, while others have concluded, at least for now. While many results are similar across the studies, significant differences emerge in various regions. Understanding the reasons behind those variances will be critical in understanding how we continue to adapt to coyotes, and how they adapt to us.

Research technician Abby-Gayle Prieur inspects the dental condition of an immobilized coyote.

What are the common patterns? In general, coyotes prefer green spaces and steer clear of highly developed areas where humans are present. Urban coyotes tend to be nocturnal, regardless of location. The urban coyote's leading cause of death is the automobile, because urban coyotes are forced to cross busy roads. Typically, urban coyotes eat natural prey, such as small rodents and rabbits, as well as fruit—and only small amounts of human-related food. Pets are rarely eaten.

As locations change, urban coyote diets may as well, a simple function of availability. Depending on the city, coyotes may be forced to use more developed areas with denser human populations. Pest control programs involving poisons may affect coyote populations, as do climate and local disease outbreaks.

Almost unique among North American cities, Los Angeles has been home to coyotes since its very beginning. Studies of coyotes in the Los Angeles area began before ours, and the initial research was less focused on "urban" animals. Yet over the years, scientists began studying more urban and suburban animals, and many similar traits to Chicago's coyotes emerged.

As in Cook County's suburbs, LA coyotes had relatively compact home ranges and preferred green spaces like parks. In more urban settings, the coyotes were more active at night and ate a diet consisting largely of natural foods. As might be expected in a car-dominated region, vehicles were a leading cause of death for coyotes.

Yet in Los Angeles, coyotes kill and eat far more cats than the coyotes in Cook County. We don't yet know why. Perhaps there are simply more free-ranging cats in the warm California climate and coyotes are taking advantage of an abundant food source? If so, it suggests that urban coyotes' diets may vary widely from city to city, with food availability playing an enormous role.

In most urban settings, coyotes are the apex predator. Not so in Southern California, where mountain lions can be seen in suburban streets and, famously, beneath the giant "Hollywood" sign in the Santa Monica foothills.

Canyons thread through the Los Angeles landscape, providing arteries of green space that coyotes and other wildlife use as travel corridors. Importantly, these canyons also are bordered by private properties, and residents often dump trash there. This creates a steady source of food for coyotes, sometimes with fatal results—coyote deaths attributed to rodent poisons are more common there.

All this may help explain why coyote attacks on people are more common in Southern California than anywhere else. The coyotes have been there longer, humans have become more used to them, and the coyotes have easy access to human food in the form of trash. Fortunately, most of the incidents are minor, but they serve as a constant reminder that coyotes are best left to themselves and that a coyote that has lost its fear of humans is a dangerous animal.

In Denver, I helped to initiate a study of urban coyotes that began in 2012. It found many consistencies with Chicago's coyotes, notably

coyotes' preferences for green spaces and small home ranges. To date, this project was the only study other than ours to use "object" testing to determine whether urban coyotes may be bolder than suburban or rural animals. Similarly, Denver's researchers tried to determine whether hazing coyotes would be effective in preventing conflicts with humans, but like us, they had only mixed results.

A study led by researchers at the University of Wisconsin–Madison continues to try to determine whether coyotes are detrimental to red foxes and how to help both species coexist with humans.

In Edmonton, Alberta, researchers used GPS collars to track coyotes. As in other studies, these coyotes are nocturnal. They found that coyotes in urban spaces were more susceptible to disease, such as mange, and that could be a function of the consumption of human-related foods. We have not yet observed this relationship between urban living and disease in the Chicago area, but it is an active area of our research.

Opposite: Over the decades of our study, tracking equipment has gotten smaller and better. Further technological advances will help us unravel many more mysteries about coyotes.

By no means are these the only ongoing or completed studies of urban coyotes. Minneapolis- and Atlanta-based scientists are cooperating with us in a multiple-city study to compare coyote boldness and shyness. Independent projects are underway in San Francisco; Long Island; Portland, Oregon; Lincoln, Nebraska; and Washington, DC; and likely many other locations.

When will the coyote experiment end? As we move forward, it is unlikely we can stop their spread, and there is every reason to believe that our relationship with the coyote will continue to evolve, ever changing, perhaps at times in response to us. Our attitudes toward coyotes will continue to evolve as well. If we allow ourselves some humility, we may learn something about ourselves.

So much of what we have learned about coyotes has been the result of advances in technology. We could learn only so much about the animals' diets until stable isotope analysis became possible, and GPS

You never know when or where a coyote might appear.

collars are a significant advance over earlier radio-tracking technology. As smaller cameras become available, as the potential life span of tracking collars lengthens, as new technologies emerge to enhance our ability to track and study these animals, we will likely learn even more about the coyotes among us and open new avenues to learn about their life on our streets.

Yet it is natural to miss the macro when focused on the micro. When we are absorbed in the fascinating details of the coyote's success in cities and become immersed in the science and technology required to uncover the many hidden layers to this amazing story, it is easy to overlook the remarkable, which is what the coyote has accomplished by colonizing cities across North America. The overarching bottom line is that coyotes can successfully reside in a city only by avoiding us and by retaining their instinct to fear us.

The coyotes' "golden rule" is to avoid people at all costs. It applies universally and has served them well across the rural areas that make up most of their range. But imagine facing that prospect in an area with millions of people, such as the Chicago metropolitan area. Most urban coyotes still adhere to that bottom line, every second of every day, as they go about their lives and we go about ours.

Going forward, science will explore the ways in which coyotes may be changing in response to urban pressures beyond the immediate changes for survival—looking, rather, to evolutionary adaptations. Are urban conditions creating a smarter animal? A bolder animal? Will coyotes be able to acquire more bold genes while retaining their ability to control aggressive behavior? Are coyotes on a path toward self-domestication?

Future research will also focus on us humans and how we are changing with respect to our attitudes and tolerance toward coyotes and other predators that force us to modify our behavior and accommodate them in our lives. For so long, it seemed that the human-coyote relationship was forever fixed in conflict, with little change over the decades except for the development of new, ever more efficient ways to kill or control coyotes.

It is my hope that our years of research have helped us understand that despite our relentless efforts, humans will be living with coyotes for as long as they share the North American continent. As such, it would be best to amend our relationship with these remarkable survivors to one of coexistence, and to view coyotes not with contempt or fear but with respect, awe, and wonder.

ACKNOWLEDGMENTS

This research has been supported financially through a unique collaboration of agency partners: Cook County Department of Animal and Rabies Control, the Forest Preserve District of Cook County, and the Max McGraw Wildlife Foundation. Other long-term partners include the Brookfield Zoo and the University of Illinois Zoological Pathology lab.

As important as the funding has been, this project would not have been possible without the support of several amazing individuals. First and foremost is Chris Anchor, senior biologist for the Forest Preserve District of Cook County. He has been a partner from the beginning, providing support in innumerable ways. This project would never have started without him and certainly would not have continued without his efforts behind the scenes to advocate for continued funding over the years. Chris provided access to locations and equipment and always supported the work in whatever way he could.

Dr. Dan Parmer, director of the Cook County Department of Animal and Rabies Control, was the person who requested a proposal from me and then agreed to fund it. He followed that up with continued support each year, increasing funding once we realized that the urban coyote phenomenon is an ongoing natural experiment. Upon

Opposite: The coyote's gaze can be unnerving in its intensity.

Dr. Parmer's death, Dr. Donna Alexander assumed his position and continued the agency's long-term commitment. Dr. Alexander also provided support beyond funding and consistently advocated for the research. Following Dr. Alexander's passing, Dr. Tom Wake became interim director and continued the support. I am indebted to each of them.

I am personally indebted to Charlie Potter, president and CEO of the Max McGraw Wildlife Foundation, who has not only supported the project but has supported my career. It was Charlie who allowed me to continue the research program with the foundation even as I professionally moved from the foundation to The Ohio State University in 2003. Equally important, Charlie continued to support the coyote research despite his own ambivalence toward certain mammalian predators that eat his beloved ducks.

The quality of our science, and our ability to better understand aspects of coyote life, benefited from collaborations with talented scientists and their institutions, including Dr. Tom Meehan of the Brookfield Zoo, Dr. Jean Dubach of Loyola University, and Dr. Seth Newsome at the University of New Mexico. I have learned so much from them.

Foundation research associates Suzie Prange, Justin Brown, Heidi Garbe, and especially Shane McKenzie supervised and trained the teams of field technicians over the years. They kept the research going and on course despite the inevitable turnover of young professionals serving as technicians each year. And, of course, thank you to those field techs, too numerous to list, who continued the data collection, often mundane work, in addition to even more mundane data entry, maintenance of equipment and vehicles, cleaning of lab equipment, and the hours and hours of driving through Chicago traffic, following the beeps in the night of largely invisible radio-collared coyotes.

ABOUT THE AUTHORS

Stanley D. Gehrt, PhD, is professor of wildlife ecology at The Ohio State University, and chair of the Center for Wildlife Research at the Max McGraw Wildlife Foundation. His research program focuses on various aspects of mammalian ecology, especially urban systems; dynamics of wildlife disease; and human-carnivore conflicts. He is principal investigator of one of the largest studies of coyotes to date: capturing and tracking more than 1,450 coyotes for the past twenty-two years in the Chicago area. His research has been featured in numerous print, radio, and television outlets, including PBS, ABC's *Nightline, NBC Nightly News,* National Geographic, and the History Channel. He lives in Columbus, Ohio.

Following page: A coyote stands guard above Los Angeles.

Kerry Luft is executive vice president of the Max McGraw Wildlife Foundation in Dundee, Illinois. He spent nearly three decades in journalism, mostly at the *Chicago Tribune,* where he specialized in national and foreign news. He has written or edited five other books, including *Wings Over Water,* the companion book to the award-winning IMAX 3D film of the same name.